# 10 Oklahoma OSTP Grade 6 Math Practice Tests

*The Ultimate Test Prep Collection with Answer Explanations*

Dr. A. Nazari

# 10 Practice Tests

## The Grand Championship Collection

*Welcome, future Math Champion!*

*You hold the **ultimate collection** —*
*ten full-length practice tests designed to take you*
*from first attempt to **complete mastery**.*

- *Conquer every Grade 6 topic*
- *Build unshakeable confidence*
- *Rise from Bronze to Gold to Champion*
- *Arrive at test day fully prepared*

*The championship begins now.*

> *Ten tests may seem like a marathon, but champions are made one step at a time. Trust the process!*

# 👑 The Champion's Path 👑

Your 4-phase journey from Bronze to Champion

## 🥉 Bronze Round (Tests 1–3)

Your warm-up matches. Take these **untimed** to learn the format and set your baseline. Read the answer explanations after each test — this is where you build your foundation.

## 🥈 Silver Round (Tests 4–6)

Set a timer for **75 minutes**. Focus on the topics that tripped you up in Bronze. Practice showing your work on every problem. Your accuracy should be climbing.

## 🥇 Gold Round (Tests 7–9)

Full timed conditions (**60 minutes**). Simulate the real exam environment. Review only the questions you missed — targeted practice is the key to gold.

## 👑 Championship Final (Test 10)

Your final match. Full exam conditions — timed, quiet, no breaks. This is your victory lap. Show yourself how far you've come!

## 🏆 Your Championship Kit

- **10 Full-Length Practice Tests** — every Grade 6 topic
- **Formula Reference Sheet**
- **Complete Answer Key** with explanations
- **Championship Scoreboard** to track your rise

 **Champion's Tip:** Space your tests 2–3 days apart. Use the days in between for targeted review. By Test 10, you'll be amazed at your transformation.

# ♚ The Champion's Playbook ♚



**I**    **Read every question twice.** The first read tells you the topic. The second tells you exactly what to solve for. Champions never skim.

**II**    **Mark the clues.** Circle key numbers, underline the question, and cross out information that's just there to distract you.

**III**    **Choose your strategy.** Before touching pencil to paper, decide: Am I setting up a ratio? Solving an equation? Finding area? Name the approach.

**IV**    **Solve, then match.** For multiple choice — work the problem on scratch paper first, then find your answer among the choices.

**V**    **Eliminate and conquer.** Cross out obviously wrong answers. If you're left with two, you've already doubled your odds. Make an educated pick.

**VI**    **Estimate to verify.** After solving, ask: "Is this answer reasonable?" A quick mental estimate catches most calculation errors.

**VII**    **Leave nothing blank.** Even a well-reasoned guess is worth more than an empty space. Use partial work to support your answer.

## 🕐 Timing Mastery

Tests 1–3: **Untimed** (build foundation)   ▷   Tests 4–6: **75 min** (build speed)   ▷   Tests 7–10: **60 min** (championship conditions)

## ★ Grade 6 Championship Topics

🏅 Ratios & Proportions    🏅 Integers & Rational Numbers    🏅 Expressions & Equations

🏅 Geometry & Measurement    🏅 Statistics & Data Analysis

*A true champion isn't someone who never makes mistakes — it's someone who learns from **every single one**. After each test, review your errors carefully. That's where the real growth happens.*



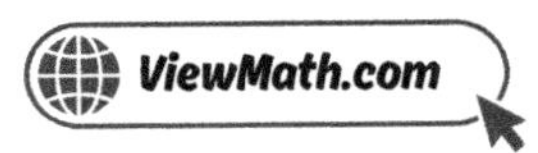

# ♛ The Champion's Toolkit ♛



## 🏆 Required Equipment

**Sharpened Pencils** — Two #2 pencils — champions always have a backup

**Quality Eraser** — A clean, soft eraser that won't smudge your work

**Scratch Paper** — Blank paper for calculations, diagrams, and number lines

**Ruler** — Essential for geometry and coordinate plane questions

**Timer** — Begin using from the Silver Round onward

**Quiet Workspace** — A calm, well-lit area free from distractions

## Permitted in Competition

- ✔ Pencil and eraser
- ✔ Scratch paper (provided)
- ✔ Ruler (if specified)
- ✔ Formula reference in this book

## Not Permitted

- ✘ Calculators
- ✘ Electronic devices
- ✘ Textbooks or notes
- ✘ Outside help

### For Parents & Teachers

- With 10 tests, space them **2–3 days apart**. This gives time to review mistakes and study between rounds.

- Let your child take Tests 1–3 untimed to build familiarity and establish a baseline.

- After each test, go through the Answer Key together. Focus on **understanding the reasoning**, not memorizing answers.

- Use the Championship Scoreboard to visualize long-term progress. Celebrate improvements at every tier!

- Pair with our **Grade 6 Math Study Guide** for topics that need sustained attention.

# Formula Reference Sheet

---

## ▲ Area Formulas

**Rectangle** $A = l \times w$

**Parallelogram** $A = b \times h$

**Triangle** $A = \dfrac{1}{2} \times b \times h$

**Trapezoid** $A = \dfrac{1}{2}(b_1 + b_2) \times h$

## ▣ Volume

**Rectangular Prism** $V = l \times w \times h$

## ▣ Surface Area

Find the area of each face, then add them all up.

**Rectangular Prism:**

$SA = 2lw + 2lh + 2wh$

## ↓ Order of Operations

**P** Parentheses first

**E** Exponents

**M/D** Multiply & Divide (left to right)

**A/S** Add & Subtract (left to right)

## ％ Ratios & Percents

**Ratio:** $a : b$ or $\dfrac{a}{b}$

**Unit rate:** amount per 1 unit

**Percent:** a ratio out of 100

Part = Percent × Whole

## ⚖ Integers & Absolute Value

Integers:

$\ldots, -3, -2, -1, 0, 1, 2, 3, \ldots$

$|-5| = 5 \qquad |5| = 5$

Absolute value = distance from 0

## X¹ Expressions & Equations

**Exponent:** $3^4 = 3 \times 3 \times 3 \times 3 = 81$

**Variable:** a letter that stands for a number

**Equation:** two expressions joined by $=$

**Inequality:** uses $<, >, \leq, \geq$

## ⊕ Coordinate Plane

Ordered pair: $(x, y)$

$x$-axis: horizontal    $y$-axis: vertical

Origin: $(0, 0)$

Four quadrants (I, II, III, IV)

## ▦ Statistics

**Mean:** sum of values $\div$ count

**Median:** middle value (sorted)

**Range:** max $-$ min

# Championship Scoreboard

Track your rise through every championship round

**Champion's Name:** ___________________________

| Round | Tier | Date | Score | Rating |
|---|---|---|---|---|
| 1 | Bronze | | | |
| 2 | Bronze | | | |
| 3 | Bronze | | | |
| 4 | Silver | | | |
| 5 | Silver | | | |
| 6 | Silver | | | |
| 7 | Gold | | | |
| 8 | Gold | | | |
| 9 | Gold | | | |
| 10 | | | | |

*My strongest topics (where I consistently score well):*

_______________________________________________

_______________________________________________

*Topics I improved on the most from Bronze to Gold:*

_______________________________________________

_______________________________________________

*My score trend (Bronze avg → Gold avg → Championship):*

_______________________________________________

_______________________________________________

*One strategy that helped me improve the most:*

_______________________________________________

_______________________________________________

*My confidence level for the real test (1–10):*  _______ / 10

Find more at
ViewMath.com/OK-Grade6

# Continue Learning at
# ViewMath Academy!

---

## For Parents, Teachers & Students

Great job on the practice tests! Want to keep improving? ViewMath Academy is your **free online companion** to this book.

- **Score Analyzer** — Enter your answers and instantly see which topics need more practice

- **Interactive Lessons** — Review the concepts behind each question with clear explanations

- **Adaptive Quizzes** — Practice your weak topics with questions that match your level

- **Progress Tracking** — See your mastery grow across all Grade 6 math topics

- **Personalized Dashboard** — A learning plan tailored just for you

**Scan to visit ViewMath Academy**

ViewMath.com/OK-Grade6

 Free to use · No downloads required · Works on any device 

#  Table of Contents

*Here's what we'll explore together!*

 *Let's learn and have fun!*

# Practice Test 1

☑ *30 Questions*

## ✏ Before You Start ✏

- ✓ **Read each question carefully** before choosing your answer.
- ✓ **Show your work** on scratch paper when you need to.
- ✓ **Skip hard questions** and come back to them later.
- ✓ **Check your answers** when you're done.
- ✓ **Take your time** — there's no rush!

★ You've Got This! ★

Do your best and show what you know!

1. Write a real-world example of a rate that uses the units "dollars" and "pounds."

*Your Answer*

2. The graph shows the relationship between hours worked and money earned.

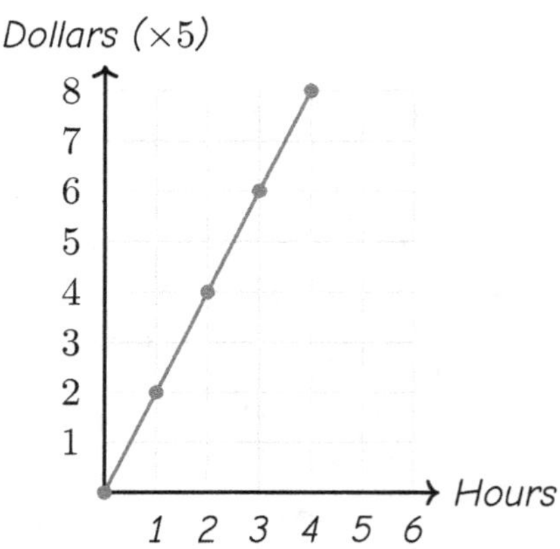


How much money is earned in 6 hours?

(A) $30

(B) $50

(C) $60

(D) $40

Find more at
ViewMath.com/OK-Grade6

ViewMath.com

3. The table and partial graph below show a ratio relationship.

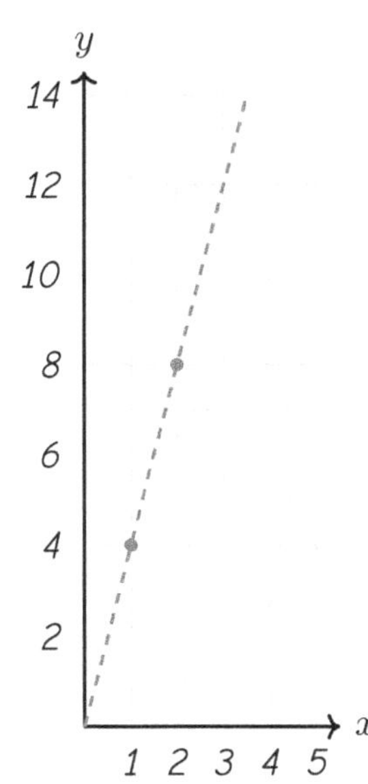

| $x$ | $y$ |
|-----|-----|
| 1 | 4 |
| 2 | 8 |
| 3 | ? |

What is the missing value of $y$ when $x = 3$?

- (A) 10
- (B) 12
- (C) 14
- (D) 16

4. What is 15% of 60?

- (A) 9
- (B) 15
- (C) 45
- (D) 6

5. The bar model below compares different length units.

| 1 yard | | |
|--------|--------|--------|
| 1 ft | 1 ft | 1 ft |

If a rope is 7 yards long, how many feet is it?

- (A) 10
- (B) 14
- (C) 21
- (D) 28

6. *A student says: "The scale is 1 inch = 4 feet. The room is 6 inches long on the blueprint, so the actual room is 1.5 feet long." Explain the error and give the correct answer.*

*Your Answer:*

7. *The number line below shows equal jumps from $0$ to $\dfrac{5}{6}$.*

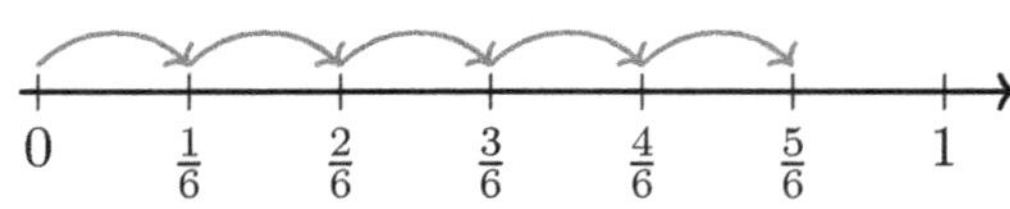

*Which division problem does the number line represent?*

(A) $\dfrac{5}{6} \div \dfrac{1}{6} = 5$

(B) $\dfrac{1}{6} \div \dfrac{5}{6} = 5$

(C) $\dfrac{5}{6} \times \dfrac{1}{6} = \dfrac{5}{36}$

(D) $\dfrac{5}{6} \div 5 = \dfrac{1}{6}$

8. *Liam reads $1{,}344$ pages in 8 weeks, reading the same number of pages each week. How many pages does he read per week?*

(A) 158

(B) 178

(C) 186

(D) 168

9. *Find $|-9| + |-6|$.*

*Your Answer:*

10. *The point $(-7, 3)$ is reflected across the $x$-axis. Write the coordinates of the reflected point.*

*Your Answer:*

11. Look at the input-output table below. Which expression was used to produce the output values?

| Input ($n$) | Output |
| --- | --- |
| 1 | 3 |
| 2 | 6 |
| 3 | 11 |
| 4 | 18 |

(A) $n^2 + 2$

(B) $3n$

(C) $n^2 + n + 1$

(D) $2n + 1$

12. Look at the two-step diagram below. Write the expression that describes the final output when starting with the input $n$.

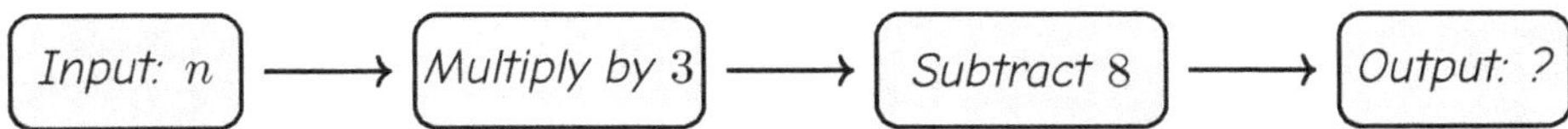


Your Answer

13. Evaluate $3(p + 2) - p$ when $p = 8$.

Your Answer

14. Which expression is equivalent to $3(x + 5)$?

(A) $3x + 5$

(B) $3x + 15$

(C) $8x$

(D) $3x + 8$

15. Write a situation that could be modeled by the expression $5n + 3$.

Your Answer

16. Solve: $\dfrac{m}{4} = 7$

(A) $m = 3$

(B) $m = 11$

(C) $m = 28$

(D) $m = 47$

17. Write a real-world situation that can be described by $x \leq 15$.

Your Answer

18. Is $x = 3$ a solution to $x \leq 3$? Is $x = 3$ a solution to $x < 3$? Explain.

Your Answer

19. Which table shows a relationship where $y$ does NOT depend on $x$?

(A) $x : 1, 2, 3 \quad y : 3, 6, 9$

(B) $x : 1, 2, 3 \quad y : 7, 7, 7$

(C) $x : 1, 2, 3 \quad y : 2, 4, 6$

(D) $x : 1, 2, 3 \quad y : 5, 10, 15$

20.  A composite figure is made of a rectangle (8 cm by 5 cm) attached to a triangle with base 8 cm and height 3 cm. What is the total area?

(A)  52 cm²                                            (B)  40 cm²

(C)  55 cm²                                            (D)  64 cm²

21.  A right triangle has vertices $(-4, 0)$, $(2, 0)$, and $(-4, 5)$. What is the area?

(A)  30 square units                                  (B)  15 square units

(C)  11 square units                                  (D)  20 square units

22.  A gift box is 12 in long, 8 in wide, and 5 in tall. How much wrapping paper is needed to cover all sides?

(A)  392 in²                                           (B)  480 in²

(C)  296 in²                                           (D)  196 in²

23.  A data set is described as "symmetric." What does this mean?

(A)  All values are the same.                          (B)  The left side and right side of the data look about the same.

(C)  The data has no center.                           (D)  There are no outliers.

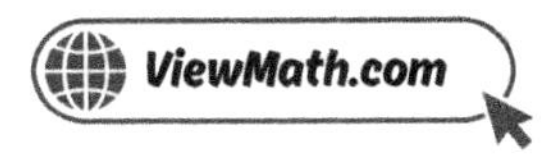

24. *Data:* $25, 30, 35, 40, 45$. *If the value* $45$ *is changed to* $100$, *what happens?*

   (A) The mean stays the same but the median changes.

   (B) The median stays the same but the mean increases.

   (C) Both the mean and median increase.

   (D) Neither changes.

25. *Shoe sizes of* $9$ *students:* $6, 7, 7, 8, 8, 8, 9, 9, 10$. *In a frequency table, which size has a frequency of* $3$?

   (A) 6

   (B) 7

   (C) 8

   (D) 9

26. *Look at the box plot below.*

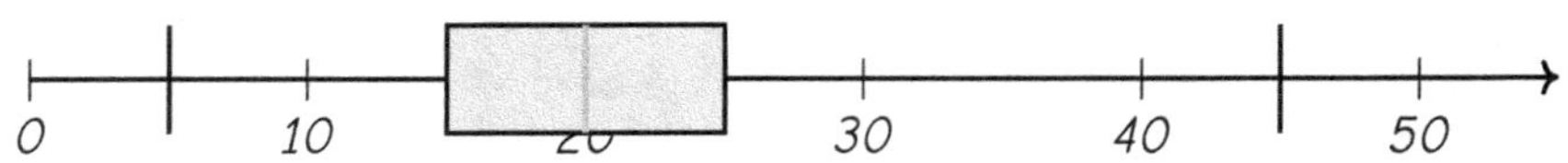


*The right whisker is much longer than the left whisker. What does this suggest about the data?*

   (A) The data is symmetric.

   (B) The data is skewed right — some high values are spread far from the box.

   (C) The data is skewed left.

   (D) The data has no outliers.

27. *Restaurant A wait times (min): median* $= 12$, *IQR* $= 4$. *Restaurant B: median* $= 10$, *IQR* $= 10$. *Which restaurant has shorter typical wait times? Which has more predictable wait times?*

   (A) B shorter, B more predictable

   (B) B shorter, A more predictable

   (C) A shorter, A more predictable

   (D) A shorter, B more predictable

28. *A bag contains* $4$ *green marbles,* $6$ *red marbles, and* $10$ *blue marbles. What is the probability of randomly drawing a green marble? Write your answer as a percent.*

   Your Answer

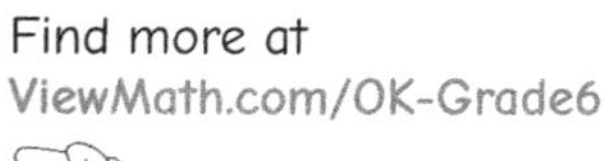
Find more at
ViewMath.com/OK-Grade6

29. *Using the stem-and-leaf plot below, how many data values are greater than 50?*

| Stem | Leaves |
|------|--------|
| 3 | 2 5 8 |
| 4 | 0 4 7 9 |
| 5 | 1 3 6 |
| 6 | 2 |

*Key:* 3 | 2 *means* 32

(A) 3

(B) 4

(C) 5

(D) 6

30. *A line graph shows a town's average temperature each month from January (28°F) to May (62°F). The values increase each month. Is this graph showing a positive trend, a negative trend, or no trend?*

Your Answer:

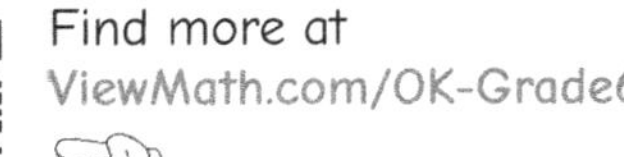


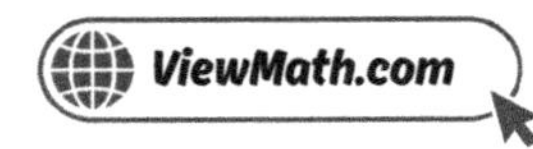

 # End of Practice Test 1 

Great job finishing the test!

 My Score

I got ______________ out of 30 questions right.

Check your answers in the **Answer Key** at the back of the book.

Review any questions you missed. That's how we learn!

## Check Your Score Online!

Visit **ViewMath Academy** to enter your answers and see which topics you need to review. You can also explore lessons, take quizzes, track your scores, and save your progress!

viewmath.com/score/6.1.OK.16

Or go to viewmath.com/score and enter code:  6.1.OK.16

2

# Practice Test 2

📋 30 Questions

✔ **Read each question carefully** before choosing your answer.

✔ **Show your work** on scratch paper when you need to.

✔ **Skip hard questions** and come back to them later.

✔ **Check your answers** when you're done.

✔ **Take your time** — there's no rush!

⭐ You've Got This! ⭐

Do your best and show what you know!

1. The bar graph shows how many cans two students collected over different numbers of days.

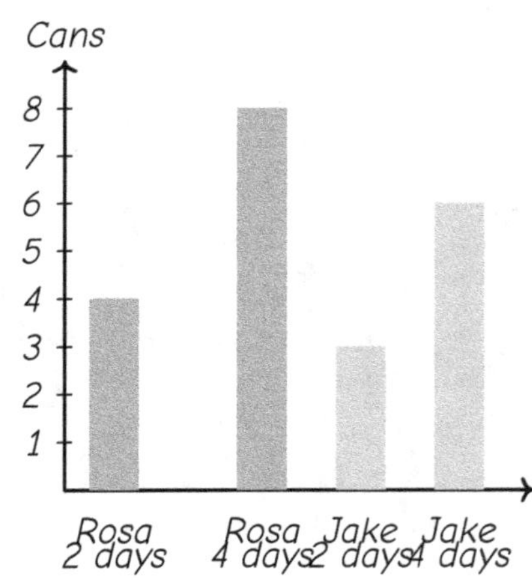


Who collects cans at a higher rate per day?

(A) Rosa, at 2 cans per day

(B) Jake, at 1.5 cans per day

(C) Rosa, at 4 cans per day

(D) They collect at the same rate.

2. Complete the ratio table below. The ratio of markers to crayons is 4 : 6.

| Markers | Crayons |
| --- | --- |
| 4 | 6 |
| 8 | ? |
| ? | 24 |
| 20 | ? |

Find all three missing values.

Your Answer

Find more at
ViewMath.com/OK-Grade6

3. Two runners' distances are graphed below.

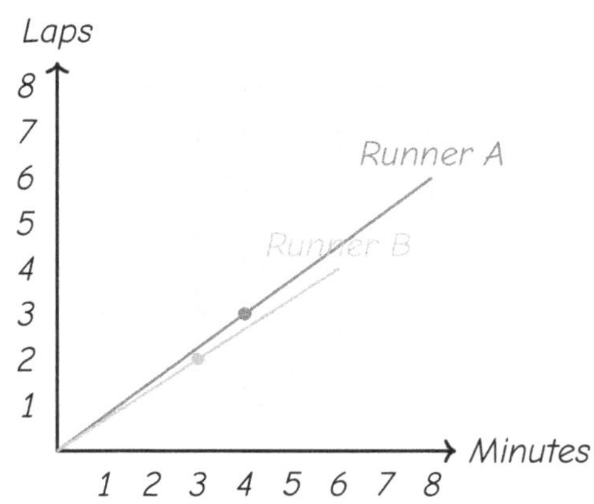


**Part A:** What is each runner's ratio of laps to minutes?

**Part B:** Who runs faster? Explain using the graph.

Your Answer

4. 25% of what number is 20?

(A) 5

(B) 40

(C) 80

(D) 100

5. A bag weighs 500 grams. What is its weight in kilograms?

(A) 50

(B) 5

(C) 0.5

(D) 0.05

6. A blueprint uses a scale of 1 inch = 3 feet. A room is 4 inches by 5 inches on the blueprint. What is the actual area of the room?

(A) 20 square feet

(B) 60 square feet

(C) 180 square feet

(D) 540 square feet

7. What is $\dfrac{3}{5} \div \dfrac{1}{5}$?

   (A) $\dfrac{3}{25}$                      (B) $3$

   (C) $\dfrac{1}{3}$                        (D) $15$

8. Compute $4{,}515 \div 15$. Be careful with every digit of the quotient.

   Your Answer:

9. What is $|-12|$?

   (A) $-12$                      (B) $12$

   (C) $0$                        (D) $-(-(-12))$

10. Explain why the point $(9, 0)$ is not in any quadrant.

    Your Answer:

11. Evaluate: $3 \times (2+4)^2$

    (A) $42$                      (B) $108$

    (C) $324$                     (D) $48$

12. Write an expression for: "a number $t$ divided by 4, then subtract 1."

    Your Answer:

13. The triangle below has a base of $b = 10$ cm and a height of $h = 6$ cm. Use the formula $A = \dfrac{1}{2} \times b \times h$ to find its area.

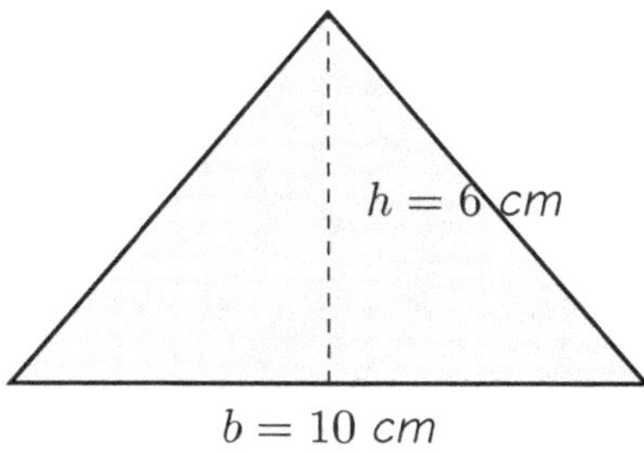


Your Answer

14. Simplify: $3a + 7 + 5a - 2a + 3$

Your Answer

15. A movie ticket costs \$9. Your group also shares a \$6 bucket of popcorn. Which expression represents the total cost for $p$ people?

  (A) $9p + 6$                  (B) $9 + 6p$

  (C) $15p$                     (D) $9p - 6$

16. Solve: $k + 15 = 32$

  (A) $k = 47$               (B) $k = 17$

  (C) $k = 27$               (D) $k = 2$

17. Which inequality represents "a number $y$ is fewer than 20"?

   (A) $y > 20$

   (B) $y \geq 20$

   (C) $y < 20$

   (D) $y \leq 20$

18. Which inequality has a graph with an open circle at 0 and shading to the right?

   (A) $x \leq 0$

   (B) $x \geq 0$

   (C) $x > 0$

   (D) $x < 0$

19. Identify the independent and dependent variables: A runner jogs 6 miles per hour. The distance $d$ depends on how many hours $h$ the runner jogs.

   Your Answer

20. A parallelogram has base 12 m and a slant side of 8 m. The height is 6 m. What is the area?

   (A) $96\ m^2$

   (B) $72\ m^2$

   (C) $48\ m^2$

   (D) $36\ m^2$

21. A rectangle on the coordinate plane has area 54 square units. Two of its vertices are at $(-3, 2)$ and $(6, 2)$. What is the width of the rectangle?

   (A) 6 units

   (B) 9 units

   (C) 3 units

   (D) 18 units

22. A classroom has dimensions 10 m by 8 m by 3 m. What is the total surface area of the walls, floor, and ceiling?

   Your Answer:

Find more at
ViewMath.com/OK-Grade6

23. *Test scores:* $68, 70, 72, 73, 74, 75, 76, 98$. *Where is the center of this data?*

(A) *Around 68–70*                              (B) *Around 73–75*

(C) *Around 83*                                 (D) *Around 98*

24. *Seven students' ages are* $11, 11, 12, 12, 12, 13, 13$. *Find the mean and median.*

Your Answer:

25. *The histogram below shows the number of pages students read last week.*

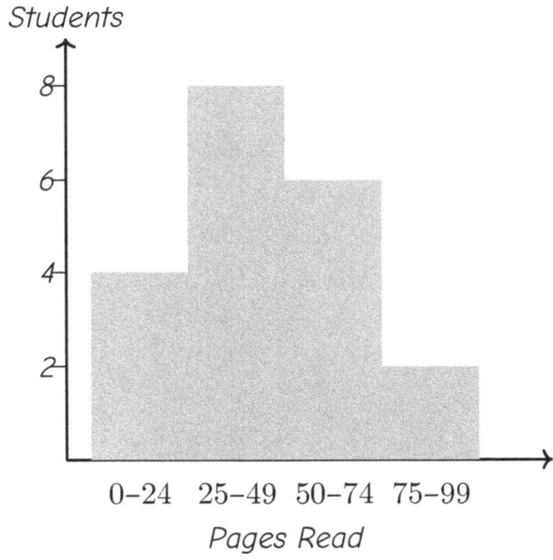


Which interval is the modal interval (contains the most students)?

(A) 0–24                                        (B) 25–49

(C) 50–74                                        (D) 75–99

26. The data table below shows the number of books 9 students read this semester.

| Student | 1 | 2 | 3 | 4 | 5 | 6 | 7 | 8 | 9 |
|---------|---|---|---|---|---|---|---|----|----|
| Books | 2 | 4 | 5 | 7 | 8 | 9 | 10 | 12 | 15 |

Find the five-number summary and the IQR.

Your Answer

27. Which two measures together best describe the "center" and "spread" of a data set?

(A) Mean and mode

(B) Median and IQR

(C) Range and maximum

(D) Minimum and maximum

28. The weather forecast says there is a 60% chance of sunshine. Which word best describes this event?

(A) Impossible

(B) Unlikely

(C) Likely

(D) Certain

29. The back-to-back stem-and-leaf plot below shows the test scores for two classes.

| Class A | Stem | Class B |
|---------|------|---------|
| 8 5 2 | 6 | 3 6 |
| 7 4 1 | 7 | 0 2 5 9 |
| 6 0 | 8 | 1 4 |
|  | 9 | 2 |

Key: 2 | 6 | 3 means 62 for Class A and 63 for Class B

Which class has the higher median score?

(A) Class A

(B) Class B

(C) They have the same median.

(D) There is not enough information to tell.

Find more at
ViewMath.com/OK-Grade6

30. The double bar graph below shows the number of books read by boys and girls in two classes. Use the graph to answer the question.

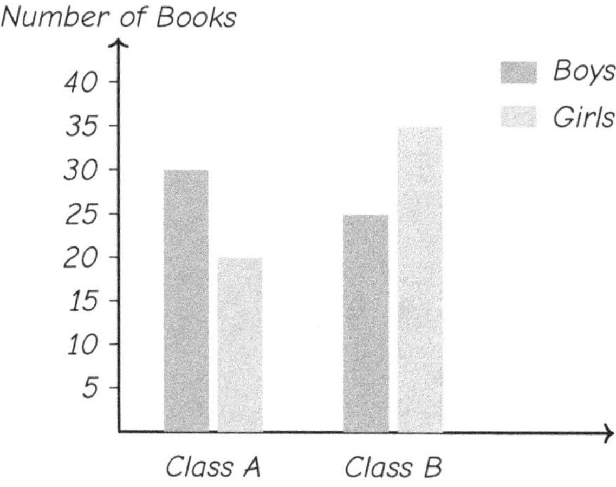


How many more books did boys in Class A read than girls in Class A?

(A) 5

(B) 10

(C) 15

(D) 20



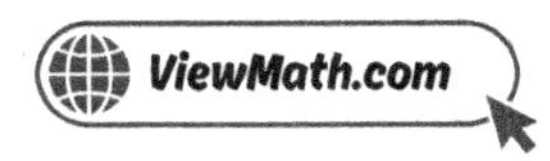

#  End of Practice Test 2 

*Great job finishing the test!*

## ☑ My Score

I got _____________ out of 30 questions right.

*Check your answers in the **Answer Key** at the back of the book.*

💡 *Review any questions you missed. That's how we learn!*

## 📊 Check Your Score Online!

Visit **ViewMath Academy** to enter your answers and see which topics you need to review. You can also explore lessons, take quizzes, track your scores, and save your progress!

viewmath.com/score/6.1.OK.17

*Or go to viewmath.com/score and enter code: 6.1.OK.17*

# Practice Test 3

📋 30 Questions

## ✏️ Before You Start ✏️

- ✓ **Read each question carefully** before choosing your answer.
- ✓ **Show your work** on scratch paper when you need to.
- ✓ **Skip hard questions** and come back to them later.
- ✓ **Check your answers** when you're done.
- ✓ **Take your time** — there's no rush!

⭐ You've Got This! ⭐

Do your best and show what you know!

1. A recipe serves 6 people and requires 2 cups of rice. How would you write the rate of people to cups of rice?

   (A) 6 cups per 2 people

   (B) 6 people per 2 cups

   (C) 2 : 6 cups

   (D) 8 total

2. A recipe uses 4 eggs for every 6 cups of flour. How many eggs are needed for 18 cups of flour?

   (A) 6

   (B) 10

   (C) 12

   (D) 8

3. A ratio graph passes through $(6, 2)$. What is the value of $y$ when $x = 15$?

   (A) 3

   (B) 5

   (C) 7

   (D) 10

4. A store sells a book for \$25. Next week the price goes up by 10%. What is the new price?

   (A) \$35

   (B) \$27.50

   (C) \$26

   (D) \$22.50

5. Convert 156 inches to feet.

   (A) 11

   (B) 12

   (C) 13

   (D) 14

Find more at
ViewMath.com/OK-Grade6

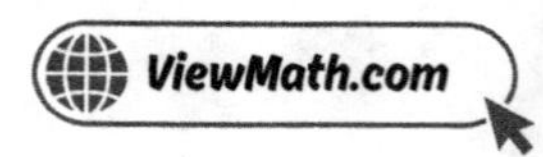

6. A map has a scale of 1 cm = 3 km. Two cities are 7 cm apart on the map. What is the actual distance between them?

(A) 10 km

(B) 21 km

(C) 30 km

(D) 42 km

7. What is $\dfrac{1}{3} \div \dfrac{2}{3}$?

(A) $\dfrac{1}{2}$

(B) $\dfrac{2}{9}$

(C) $\dfrac{2}{3}$

(D) 2

8. What is $1{,}575 \div 5$?

(A) 305

(B) 3,150

(C) 315

(D) 351

9. Which number has an absolute value of 6?

(A) 6 only

(B) −6 only

(C) Both 6 and −6

(D) No number has an absolute value of 6

10. Which sign convention describes all points in Quadrant II?

(A) $(+,+)$

(B) $(-,+)$

(C) $(-,-)$

(D) $(+,-)$

11. Which shows $5 \times 5 \times 5$ written using an exponent?

$A$  $5 + 3$                                          $B$  $5 \times 3$

$C$  $3^5$                                            $D$  $5^3$

12. Write an expression for: "the sum of a and b, divided by 2."

Your Answer:

13. Evaluate $k^2 + k$ when $k = 3$.

$A$  6                                                $B$  9

$C$  12                                               $D$  15

14. Use the distributive property to expand $8(y + 3)$.

Your Answer:

15. In the formula $d = 60t$, what could d and t represent?

$A$  d = days, t = hours                              $B$  d = distance in miles, t = time in hours

$C$  d = dollars, t = tax rate                        $D$  d = degrees, t = temperature

16. Solve: $x + 24 = 50$

$A$  $x = 26$                                          $B$  $x = 74$

$C$  $x = 36$                                          $D$  $x = 2$

Find more at
ViewMath.com/OK-Grade6

ViewMath.com

17. The table below matches phrases to inequality symbols. Which row has an error?

| Row | Phrase | Symbol |
|---|---|---|
| 1 | At least | $\geq$ |
| 2 | No more than | $\leq$ |
| 3 | Fewer than | $<$ |
| 4 | At most | $<$ |

(A) Row 1

(B) Row 2

(C) Row 3

(D) Row 4

18. The pool opens when the temperature is more than 75°F. Which describes the graph of this inequality?

(A) Closed circle at 75, shade right

(B) Open circle at 75, shade right

(C) Closed circle at 75, shade left

(D) Open circle at 75, shade left

19. A recipe uses 2 cups of flour for every batch of cookies. If you make $b$ batches, which equation gives the total flour $f$?

(A) $f = b + 2$

(B) $f = 2b$

(C) $b = 2f$

(D) $f = b \div 2$

20. A park is shaped like a trapezoid with parallel sides of 40 m and 60 m and a height of 30 m. What is the area of the park?

(A) $1,800 \ m^2$

(B) $1,200 \ m^2$

(C) $2,400 \ m^2$

(D) $1,500 \ m^2$

21. A rectangle has vertices $(0, 0)$, $(7, 0)$, $(7, 3)$, and $(0, 3)$. A right triangle is cut from one corner with legs of 2 and 3. What is the area of the remaining shape?

(A) 18 square units

(B) 21 square units

(C) 24 square units

(D) 15 square units

22. The net below shows a triangular prism. The triangles are right triangles with legs 3 cm and 4 cm. The three rectangles have widths matching the triangle sides and a shared length of 8 cm. Find the total surface area.

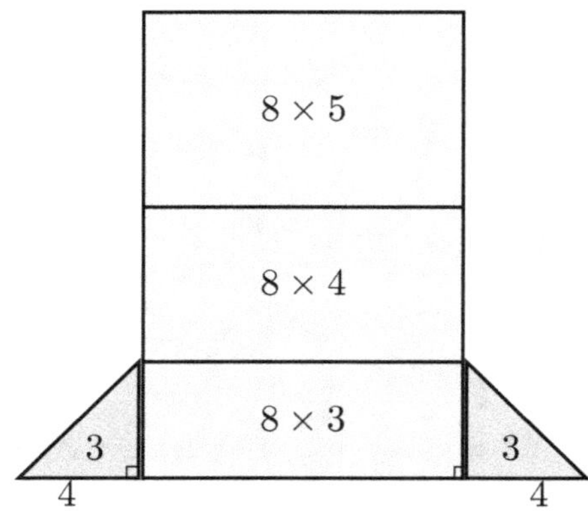


Your Answer:

23. Data set A: $20, 21, 22, 23, 24$. Data set B: $10, 15, 22, 29, 34$. Both have a center near 22. What is different?

(A) Data set A is more spread out.

(B) Data set B is more spread out.

(C) Both have the same spread.

(D) Neither data set has a center.

24. The mean of five numbers is 20. What is the sum of the five numbers?

(A) 4

(B) 25

(C) 80

(D) 100

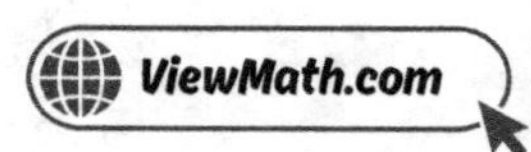

25. *A data set is:* $5, 5, 5, 5, 5$. *What is the mode?*

(A) 0

(B) 1

(C) 5

(D) *There is no mode.*

26. *A box plot has:* $min = 60$, $Q1 = 70$, $median = 75$, $Q3 = 85$, $max = 100$. *What percent of data falls between* 70 *and* 85?

27. *Data set X:* $10, 20, 30, 40, 50, 60, 70$. *Data set Y:* $35, 37, 38, 40, 42, 43, 45$. *Both have median* 40. *Which data set has a smaller IQR?*

(A) *Data set X*

(B) *Data set Y*

(C) *They are the same.*

(D) *Cannot be determined.*

28. *A fair coin is flipped once. What is the probability of landing on heads?*

(A) $\dfrac{1}{4}$

(B) $\dfrac{1}{3}$

(C) $\dfrac{1}{2}$

(D) 1



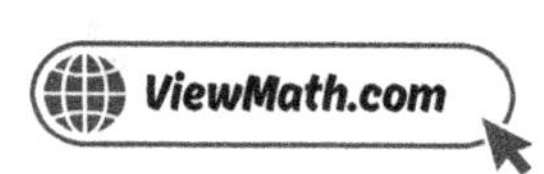

29. *What is the range of the data shown in the stem-and-leaf plot?*

| Stem | Leaves |
|------|--------|
| 2 | 3 6 |
| 3 | 0 4 7 |
| 4 | 1 5 8 |
| 5 | 2 |

*Key:* 2 | 3 *means* 23

(A) 3

(B) 29

(C) 30

(D) 52

30. *A frequency table records the number of pets owned by 25 students. The frequencies are: 0 pets (6), 1 pet (10), 2 pets (5), 3 pets (4). How many students own at least 2 pets?*

(A) 5

(B) 9

(C) 4

(D) 15

Find more at
ViewMath.com/OK-Grade6

ViewMath.com

 # ★ End of Practice Test 3 ★ 

Great job finishing the test!

##  My Score

I got ___________ out of 30 questions right.

Check your answers in the **Answer Key** at the back of the book.

💡 Review any questions you missed. That's how we learn!

### 📊 Check Your Score Online!

Visit **ViewMath Academy** to enter your answers and see which topics you need to review. You can also explore lessons, take quizzes, track your scores, and save your progress!

viewmath.com/score/6.1.OK.18

Or go to viewmath.com/score and enter code: 6.1.OK.18

# Practice Test 4

 30 Questions

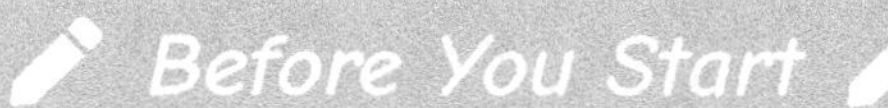

✏️ **Before You Start** ✏️

- ✓ **Read each question carefully** before choosing your answer.
- ✓ **Show your work** on scratch paper when you need to.
- ✓ **Skip hard questions** and come back to them later.
- ✓ **Check your answers** when you're done.
- ✓ **Take your time** — there's no rush!

 **You've Got This!** 

*Do your best and show what you know!*

1. Which of the following is a rate?

   (A) 5 red marbles to 3 blue marbles

   (B) 60 *miles per hour*

   (C) 4 boys to 6 gir's

   (D) 7 cats to 3 dogs

2. Ella mixes paint in a ratio of 3 parts red to 5 parts white. She uses 24 parts of red. How many total parts of paint does she have?

   *Your Answer*

3. A line passes through the origin and the point $(5, 15)$. What is the ratio $x$ to $y$?

   (A) $1 : 5$

   (B) $5 : 1$

   (C) $1 : 3$

   (D) $3 : 1$

4. A schocl has 500 students. 72% passed the state exam. How many students passed?

   *Your Answer*

5. How many inches are in 5 feet?

   (A) 50

   (B) 55

   (C) 60

   (D) 72

6. A blueprint uses a scale of 1 inch $=$ 4 feet. A wall is 20 feet long. How long is the wall on the blueprint?

   (A) 4 *inches*

   (B) 5 *inches*

   (C) 16 *inches*

   (D) 80 *inches*

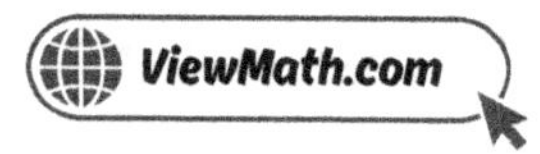

7. If $\dfrac{a}{b} \div \dfrac{c}{d} = \dfrac{a}{b} \times \dfrac{?}{?}$, what fraction replaces the question marks?

(A) $\dfrac{d}{c}$

(B) $\dfrac{c}{d}$

(C) $\dfrac{b}{a}$

(D) $\dfrac{a}{b}$

8. A factory produces 5,376 bottles and packs them into cases of 32. How many cases are needed?

Your Answer:

9. A diver is at $-40$ feet and a bird is at 25 feet above sea level. Who is farther from sea level?

(A) The bird, because $25 > -40$

(B) They are the same distance from sea level

(C) The diver, because $|-40| > |25|$

(D) Neither, because you cannot compare positive and negative numbers

10. In which quadrant is the point $(-4, 7)$ located?

(A) Quadrant I

(B) Quadrant II

(C) Quadrant III

(D) Quadrant IV

11. Evaluate: $20 - 4 \times 3 + 1$

(A) 9

(B) 49

(C) 51

(D) 7

12. A notebook costs \$3. Write an expression for the cost of $n$ notebooks plus a \$2 pen.

Your Answer:

13. *Look at the formula and the rectangle below. What is the perimeter of the rectangle?*

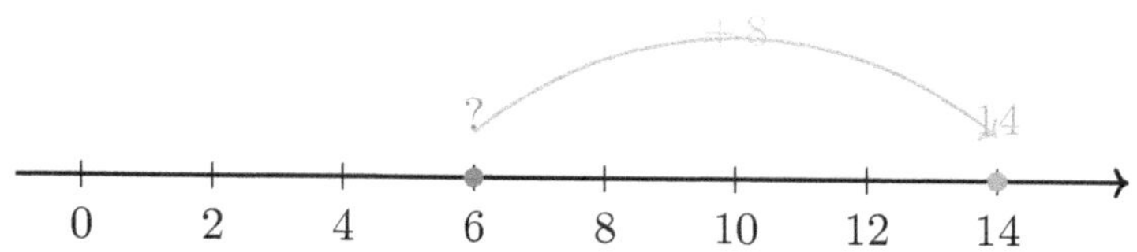


(A) 10 cm

(B) 20 cm

(C) 21 cm

(D) 42 cm

14. *Simplify:* $2(4m + 1) + 3m$

15. *In the formula* $P = 4s$, *the variable* $s$ *represents the side length of a square. What does* $P$ *represent?*

(A) The area of the square

(B) The perimeter of the square

(C) The number of sides

(D) The diagonal of the square

16. *The number line below shows a jump. Which equation does this model represent?*



(A) $x + 8 = 14$, *so* $x = 6$

(B) $x - 8 = 14$, *so* $x = 22$

(C) $8x = 14$, *so* $x = 1.75$

(D) $x + 14 = 8$, *so* $x = -6$

17. Which of the following values is NOT a solution to $y \geq 3$?

(A) $y = 3$                                (B) $y = 4$

(C) $y = 100$                              (D) $y = 2.5$

18. Describe the graph of $n \leq 5$.

(A) Open circle at 5, shade left          (B) Closed circle at 5, shade left

(C) Open circle at 5, shade right         (D) Closed circle at 5, shade right

19. The number of inches $i$ equals 12 times the number of feet $f$: $i = 12f$. How many inches are in 5 feet?

(A) 17 inches                             (B) 48 inches

(C) 60 inches                             (D) 125 inches

20. In a parallelogram, which measurement is used as the height?

(A) The slanted side                      (B) The diagonal

(C) The perpendicular distance between the base     (D) The longest side
and its opposite side

21. A right triangle has vertices $(2, 1)$, $(2, 7)$, and $(8, 1)$. What is the area?

(A) 36 square units                       (B) 18 square units

(C) 12 square units                       (D) 24 square units

Find more at
ViewMath.com/OK-Grade6

**ViewMath.com**

22. A triangular prism has 2 triangular bases and 3 rectangular faces. How many faces does it have in total?

(A) 3

(B) 4

(C) 5

(D) 6

23. Look at the dot plot below.

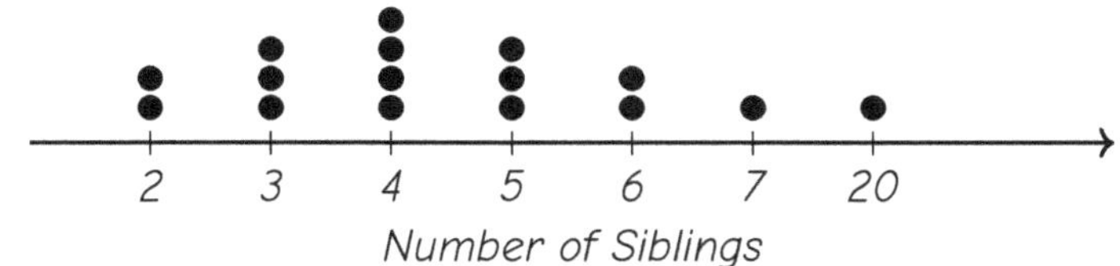


Which value is an outlier?

(A) 2

(B) 4

(C) 7

(D) 20

24. Data: $3, 4, 5, 6, 7, 8, 60$. Find the mean and median. Which better describes a typical value?

Your Answer

25. Which type of data display organizes data into a list of values (or categories) and their counts?

(A) Box plot

(B) Frequency table

(C) Dot plot

(D) Circle graph

26. A box plot shows min $= 30$, $Q1 = 45$, median $= 50$, $Q3 = 60$, max $= 90$. What percent of the data is between 45 and 90?

(A) 25%

(B) 50%

(C) 75%

(D) 100%

Find more at
ViewMath.com/OK-Grade6

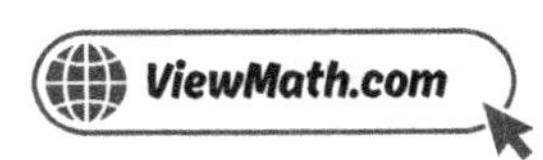

27. Which situation would make the mean a poor measure of center?

(A) A data set with all equal values

(B) A data set that is perfectly symmetric

(C) A data set with one value much larger than the rest

(D) A data set with an even number of values

28. The probability that it will rain tomorrow is 0.3. What is the probability that it will **not** rain tomorrow?

(A) 0.3

(B) 0.7

(C) 0.03

(D) 1.3

29. The back-to-back stem-and-leaf plot below shows the ages of members in two book clubs. What is the range of ages for Group X?

| Group X | Stem | Group Y |
|---|---|---|
| 9 5 2 | 2 | 1 3 7 |
| 8 6 3 | 3 | 0 4 9 |
| 1 | 4 | 2 5 |

Key: 2 | 2 | 1 means 22 for Group X and 21 for Group Y

Your Answer:

30. On a line graph, the line goes steeply upward from April to May and then stays flat from May to June. What does this tell you?

(A) The value decreased from April to June

(B) The value increased sharply, then stayed the same

(C) The value stayed the same the entire time

(D) The value increased at a steady rate

Find more at
ViewMath.com/OK-Grade6

 # ⭐ End of Practice Test 4 ⭐ 

Great job finishing the test!

##  My Score

I got _____________ out of 30 questions right.

Check your answers in the **Answer Key** at the back of the book.

💡 Review any questions you missed. That's how we learn!

### 📊 Check Your Score Online!

Visit **ViewMath Academy** to enter your answers and see which topics
you need to review. You can also explore lessons, take quizzes, track
your scores, and save your progress!

viewmath.com/score/6.1.OK.19

Or go to viewmath.com/score and enter code: 6.1.OK.19

# Practice Test 5

 30 Questions

## ✏️ Before You Start ✏️

- ✓ **Read each question carefully** before choosing your answer.
- ✓ **Show your work** on scratch paper when you need to.
- ✓ **Skip hard questions** and come back to them later.
- ✓ **Check your answers** when you're done.
- ✓ **Take your time** — there's no rush!

 ⭐ You've Got This! ⭐ 

Do your best and show what you know!

1. A garden hose fills a 300-gallon tank in 6 hours. What is the rate in gallons per hour?

Your Answer

2. Which pair of ratios are equivalent?

(A) $2 : 3$ and $4 : 9$

(B) $2 : 3$ and $6 : 9$

(C) $2 : 3$ and $8 : 9$

(D) $2 : 3$ and $3 : 2$

3. A graph of equivalent ratios passes through $(0, 0)$ and $(3, 2)$. Which statement is true?

(A) The point $(9, 4)$ is on the line.

(B) The point $(6, 4)$ is on the line.

(C) The point $(6, 5)$ is on the line.

(D) The point $(9, 8)$ is on the line.

4. The table shows the original prices and discount percents for three items.

| Item | Original Price | Discount |
|------|----------------|----------|
| Backpack | $40 | 15% |
| Shoes | $60 | 10% |
| Hat | $20 | 25% |

Which item has the greatest dollar amount of savings?

(A) Backpack ($6.00)

(B) Shoes ($6.00)

(C) Hat ($5.00)

(D) Backpack and shoes are tied.

5. How many meters are in 4.5 kilometers?

(A) 45

(B) 450

(C) 4,500

(D) 45,000

Find more at
ViewMath.com/OK-Grade6

6. A blueprint uses a scale of 1 inch = 5 feet. A bedroom on the blueprint is 3 inches by 2.5 inches. What is the actual area of the bedroom?

Your Answer:

7. What is $\dfrac{2}{3} \div \dfrac{4}{5}$?

(A) $\dfrac{8}{15}$

(B) $\dfrac{6}{5}$

(C) $\dfrac{5}{6}$

(D) $\dfrac{2}{15}$

8. In the standard long-division algorithm, the four repeating steps in order are: Divide, _________, Subtract, Bring Down. What is the missing step?

(A) Estimate

(B) Multiply

(C) Check

(D) Add

9. Which statement is **true**?

(A) $|-10| = -10$

(B) $|-10| = 10$

(C) $|-10| = 0$

(D) $|-10| = -(-(-10))$

10. Starting at the origin, you move 4 units to the left and 3 units up, then 7 units to the right and 5 units down. What point do you end at?

Your Answer:

11. *Evaluate:* $7^2 - 10$

*Your Answer*

12. *Which expression represents "the quotient of $m$ and 5"?*

(A) $5m$                                    (B) $m + 5$

(C) $m \div 5$                              (D) $5 \div m$

13. *Evaluate $m^2 + 2m + 1$ when $m = 4$*

*Your Answer*

14. *Simplify:* $4(3y - 2)$

(A) $12y - 2$                               (B) $12y - 8$

(C) $7y - 2$                                (D) $7y - 6$

15. *A store sells notebooks for \$3 each and pens for \$1 each. Which expression gives the total cost for $n$ notebooks and $p$ pens?*

(A) $3 + n + 1 + p$                         (B) $3n + p$

(C) $4np$                                   (D) $n + 3p$

16. *Solve:* $\dfrac{w}{8} = 9$

*Your Answer*

17. Which phrase matches the inequality $t \leq 30$?

(A) The temperature is more than 30 degrees

(B) The temperature is exactly 30 degrees

(C) The temperature is no more than 30 degrees

(D) The temperature is at least 30 degrees

18. Which inequality matches a graph with a closed circle at 3 and shading to the right?

(A) $x > 3$

(B) $x < 3$

(C) $x \geq 3$

(D) $x \leq 3$

19. In the equation $y = x + 4$, which statement is true?

(A) $x$ is the dependent variable

(B) $y$ is always 4

(C) $x$ is the independent variable

(D) $y$ decreases as $x$ increases

20. A trapezoid has bases $b_1 = 4$ cm and $b_2 = 4$ cm with height 6 cm. What shape is this trapezoid actually equivalent to?

(A) A triangle

(B) A parallelogram

(C) A circle

(D) A pentagon

21. An L-shaped figure has vertices $(0,0)$, $(10,0)$, $(10,4)$, $(6,4)$, $(6,8)$, and $(0,8)$. What is the total area?

Your Answer:

22. A rectangular prism has dimensions 10 m, 3 m, and 7 m. What is its surface area?

(A) $210\ m^2$

(B) $242\ m^2$

(C) $262\ m^2$

(D) $121\ m^2$

Find more at
ViewMath.com/OK-Grade6

23. Which of these is NOT a way to describe a data set?

(A) Center

(B) Spread

(C) Shape

(D) Color

24. What is the mean of the data set $4, 6, 8, 10, 12$?

(A) 6

(B) 8

(C) 10

(D) 12

25. A dot plot shows quiz scores: 6 (2 dots), 7 (5 dots), 8 (4 dots), 9 (3 dots), 10 (1 dot). What is the mode?

(A) 6

(B) 7

(C) 8

(D) 10

26. Data: $20, 25, 30, 35, 40, 45, 50, 55, 60$. What is Q3?

(A) 40

(B) 45

(C) 50

(D) 52.5

27. A student says "The data set with the larger range is always more spread out than the one with the smaller range." Is this always true? Explain.

Your Answer:

Find more at
ViewMath.com/OK-Grade6

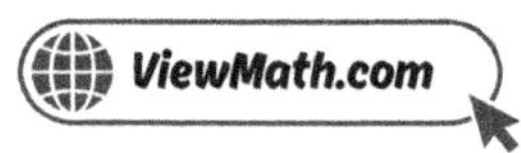

28. A standard number cube is rolled. What is the probability of rolling a 5, expressed as a decimal rounded to the nearest hundredth?

(A) 0.50

(B) 0.20

(C) 0.17

(D) 0.83

29. Use the stem-and-leaf plot to find the mean of the data.

| Stem | Leaves |
|---|---|
| 1 | 2 8 |
| 2 | 0 4 6 |
| 3 | 0 2 8 |

Key: 1 | 2 means 12

Your Answer

30. A bar graph is missing its title. The horizontal axis is labeled "Favorite Sport" and the vertical axis is labeled "Number of Students." Which title best fits this graph?

(A) Students' Heights in Sixth Grade

(B) Favorite Sports of Sixth Graders

(C) Monthly Rainfall in Our City

(D) Temperature Over One Week

Find more at
ViewMath.com/OK-Grade6

#  End of Practice Test 5 

## Great job finishing the test!

### My Score

I got _____________ out of 30 questions right.

Check your answers in the **Answer Key** at the back of the book.

Review any questions you missed. That's how we learn!

### Check Your Score Online!

Visit **ViewMath Academy** to enter your answers and see which topics you need to review. You can also explore lessons, take quizzes, track your scores, and save your progress!

viewmath.com/score/6.1.OK.20

Or go to viewmath.com/score and enter code: 6.1.OK.20

# Practice Test 6

 *30 Questions*

##  Before You Start

- ✔ **Read each question carefully** before choosing your answer.
- ✔ **Show your work** on scratch paper when you need to.
- ✔ **Skip hard questions** and come back to them later.
- ✔ **Check your answers** when you're done.
- ✔ **Take your time** — there's no rush!

 You've Got This! 

*Do your best and show what you know!*

1. A cyclist rides 42 miles in 3 hours. Write this as a rate with units.

Your Answer

2. Look at this ratio table. What is the missing value?

| Apples | Oranges |
| --- | --- |
| 3 | 5 |
| 6 | ? |

(A) 8

(B) 10

(C) 15

(D) 11

3. Which table matches a graph that passes through $(0,0)$, $(2,3)$, and $(4,6)$?

(A)

| $x$ | $y$ |
| --- | --- |
| 2 | 3 |
| 6 | 9 |

(B)

| $x$ | $y$ |
| --- | --- |
| 2 | 3 |
| 6 | 8 |

(C)

| $x$ | $y$ |
| --- | --- |
| 3 | 2 |
| 6 | 9 |

(D)

| $x$ | $y$ |
| --- | --- |
| 2 | 4 |
| 6 | 12 |

4. What is 10% of 350?

(A) 3.5

(B) 35

(C) 350

(D) 70

5. A race is 10 kilometers. How many centimeters is that?

   (A) 100,000

   (B) 10,000

   (C) 1,000,000

   (D) 1,000

6. A model car has a scale of 1:24. What does this mean?

   (A) The model is 24 times larger than the real car

   (B) Every 1 cm on the model represents 24 cm on the real car

   (C) The real car is 24 cm long

   (D) The model and the car are the same size

7. A recipe uses $\frac{2}{3}$ cup of sugar per batch. You have 4 cups of sugar. How many batches can you make?

   (A) $\frac{8}{3}$

   (B) 2

   (C) 6

   (D) 12

8. A student divides $3{,}612 \div 12$ and gets 31. The correct answer is 301. What mistake did the student most likely make?

   (A) The student forgot to subtract

   (B) The student skipped placing a zero in the quotient

   (C) The student used the wrong divisor

   (D) The student multiplied instead of dividing

9. What is the opposite of $-3$?

   (A) $-3$

   (B) 0

   (C) $\frac{1}{3}$

   (D) 3

Find more at
ViewMath.com/OK-Grade6

10. *Starting at the origin, how do you plot the point $(-3, 4)$?*

(A) *Move 3 units right and 4 units up*
(B) *Move 3 units left and 4 units down*

(C) *Move 3 units left and 4 units up*
(D) *Move 4 units left and 3 units up*

11. *Evaluate:* $2^3 + 3^2$

(A) 13
(B) 17

(C) 25
(D) 36

12. *Look at the model below. Each box represents a value. Which expression matches the model?*

$$\boxed{x}\ \boxed{x}\ \boxed{x}\ +\ \boxed{1}\ \boxed{1}$$

(A) $3x + 2$
(B) $2x + 3$

(C) $5x$
(D) $3 + 2x$

13. *Evaluate* $\dfrac{x^2}{4}$ *when* $x = 8$.

Your Answer

14. *Simplify:* $3(x + 4) + 2x$

(A) $5x + 4$
(B) $5x + 12$

(C) $6x + 12$
(D) $3x + 6$

Find more at
ViewMath.com/OK-Grade6

ViewMath.com

15. Emma earns \$8 per hour babysitting. She also got a \$15 tip. Which expression gives her total earnings for $h$ hours?

A) $8 + 15h$

B) $23h$

C) $8h + 15$

D) $8h - 15$

16. Solve: $x + 3.5 = 10$

A) $x = 6.5$

B) $x = 13.5$

C) $x = 7.5$

D) $x = 3.5$

17. Write an inequality: A suitcase must weigh less than 50 pounds. Let $w$ = weight.

Your Answer:

18. Pick two values that ARE solutions to $x > -1$ and one value that is NOT.

Your Answer:

19. A car wash charges \$10 per car. Which variable goes on the $x$-axis when graphing?

A) Total cost

B) Number of cars

C) Price per car

D) Profit

20. A kite-shaped sign can be split into two triangles, each with base 6 in and height 8 in. What is the total area of the sign?

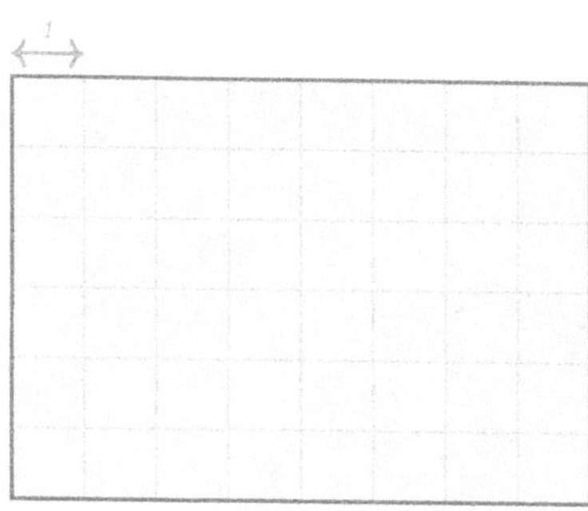

Your Answer:

21. A rectangle on the coordinate plane has an area of 56 square units. Its length is 8 units. What is the width?

(A) 6 units

(B) 7 units

(C) 8 units

(D) 48 units

22. A box is 15 in long, 6 in wide, and 4 in tall. How many square inches of cardboard are needed to make the box?

Your Answer:

23. Which describes the data set 50, 50, 50, 50, 50?

(A) Symmetric with a large spread

(B) Symmetric with zero spread

(C) Skewed right

(D) Has an outlier

24. Data: $2, 3, 4, 5, 100$. The mean is 22.8. Why is the mean much higher than most of the values?

   (A) There are only 5 values.

   (B) The median is too low.

   (C) The outlier 100 pulls the mean up.

   (D) The values are all positive.

25. A data set has two values that each appear 4 times, and no other value appears more than 3 times. How many modes does this data set have?

   (A) 0

   (B) 1

   (C) 2

   (D) 4

26. Two box plots have the same median. Class A has IQR $= 8$ and Class B has IQR $= 24$. What can you conclude?

   (A) Class A and Class B performed equally.

   (B) Class A's scores are more spread out.

   (C) Class B's middle 50% of scores are more spread out.

   (D) The range of Class A is larger.

27. A coach records sprint times (seconds) for two athletes over 10 races. Athlete A: median $= 12.5$, range $= 1.2$. Athlete B: median $= 12.5$, range $= 3.0$. Which athlete is more consistent?

   (A) Athlete A

   (B) Athlete B

   (C) They are equally consistent.

   (D) Cannot be determined without the IQR.

28. Which word best describes an event with a probability of $\frac{1}{10}$?

   (A) Certain

   (B) Likely

   (C) Equally likely

   (D) Unlikely

29. How many data values are shown in the stem-and-leaf plot below?

| Stem | Leaves |
|------|--------|
| 1 | 2 5 |
| 2 | 0 3 7 |
| 3 | 1 4 4 9 |
| 4 | 6 |

*Key:* 1 | 2 *means* 12

(A) 4

(B) 9

(C) 10

(D) 12

30. A frequency table shows that 5 students chose pizza, 8 chose tacos, 3 chose pasta, and 4 chose salad. How many students were surveyed in all?

(A) 16

(B) 18

(C) 20

(D) 24



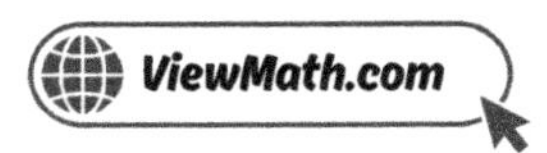

 # End of Practice Test 6 

Great job finishing the test!

 **My Score**

I got _____________ out of 30 questions right.

Check your answers in the **Answer Key** at the back of the book.

💡 Review any questions you missed. That's how we learn!

📊 **Check Your Score Online!**

Visit **ViewMath Academy** to enter your answers and see which topics you need to review. You can also explore lessons, take quizzes, track your scores, and save your progress!

viewmath.com/score/6.1.OK.21

Or go to viewmath.com/score and enter code:  6.1.OK.21

# Practice Test 7

30 Questions

---

## ✏️ Before You Start ✏️

- ✓ **Read each question carefully** before choosing your answer.
- ✓ **Show your work** on scratch paper when you need to.
- ✓ **Skip hard questions** and come back to them later.
- ✓ **Check your answers** when you're done.
- ✓ **Take your time** — there's no rush!

⭐ You've Got This! ⭐

Do your best and show what you know!

1. Is "7 pencils to 3 pencils" a rate? Explain why or why not.

   *Your Answer:*

2. A ratio table for pens to notebooks starts with 2 : 3. Which of these is a valid row?

   (A)  4 : 5
   (B)  8 : 12
   (C)  6 : 12
   (D)  10 : 12

3. A ratio graph passes through $(4, 10)$ and the origin. List two other points on this line.

   *Your Answer:*

4. A town has 4,000 residents. 15% are children. How many children live in the town?

   (A)  60
   (B)  600
   (C)  150
   (D)  1,500

5. A recipe calls for 3 pints of broth. You have a 1-quart container. How many quarts is 3 pints? (1 quart = 2 pints) Will one container be enough?

   *Your Answer:*

6. Which best describes a scale drawing?

   (A)  A drawing that is always the same size as the real object
   (B)  A drawing that is proportionally smaller or larger than the real object
   (C)  A drawing that shows only the outline of an object
   (D)  A drawing that uses no measurements

Find more at
ViewMath.com/OK-Grade6

7. A carpenter has $2\frac{1}{2}$ feet of wood. Each shelf needs $\frac{5}{8}$ of a foot. How many shelves can the carpenter cut?

Your Answer

8. What is $4{,}752 \div 12$?

(A) 396

(B) 369

(C) 406

(D) 39 R6

9. What is the opposite of 5?

(A) 5

(B) $-5$

(C) 0

(D) $\frac{1}{5}$

10. Where is the point $(-6, 0)$ located on the coordinate plane?

(A) On the $y$-axis

(B) On the $x$-axis

(C) In Quadrant II

(D) In Quadrant III

11. Evaluate: $5^2 + 4^2$

Your Answer

12. Sam has $x$ stickers. He gives away 6. Which expression shows how many stickers Sam has now?

(A) $x + 6$

(B) $6x$

(C) $6 - x$

(D) $x - 6$

Find more at
ViewMath.com/OK-Grade6

ViewMath.com

13. The perimeter of a rectangle is $P = 2l + 2w$. What is $P$ when $l = 9$ and $w = 4$?

(A) 13

(B) 22

(C) 26

(D) 36

14. Simplify: $5p + 3 + 2p + 4p - 1$

(A) $11p + 2$

(B) $7p + 2$

(C) $11p + 4$

(D) $12p$

15. A taxi charges \$3 plus \$2 per mile. Write an expression for the cost of a ride that is $m$ miles long.

Your Answer:

16. Solve: $5n = 45$

(A) $n = 9$

(B) $n = 40$

(C) $n = 50$

(D) $n = 225$

17. Which inequality represents "a number $x$ is greater than 7"?

(A) $x < 7$

(B) $x > 7$

(C) $x \leq 7$

(D) $x = 7$

18. A student graphs $x > 2$ with a closed circle at 2 and shading to the right. What is the error?

(A) Should shade to the left

(B) Should use an open circle at 2

(C) Should use a circle at 0 instead

(D) There is no error

Get Online

Find more at
ViewMath.com/OK-Grade6

ViewMath.com

19. A phone battery loses 5% each hour. Which is the independent variable?

(A) Battery percentage

(B) Phone model

(C) Number of hours

(D) Screen brightness

20. A trapezoid has an area of 84 cm$^2$ and bases of 10 cm and 14 cm. What is the height?

Your Answer

21. An L-shaped figure has vertices $(0,0)$, $(8,0)$, $(8,3)$, $(4,3)$, $(4,7)$, and $(0,7)$. A student splits it into two rectangles. Which pair of rectangles correctly covers the figure?

(A) $8 \times 7$ and $4 \times 3$

(B) $8 \times 3$ and $4 \times 4$

(C) $4 \times 7$ and $4 \times 3$

(D) $8 \times 3$ and $4 \times 7$

22. A rectangular prism has dimensions 4 cm by 4 cm by 9 cm. What is the surface area?

(A) $144 \ cm^2$

(B) $176 \ cm^2$

(C) $160 \ cm^2$

(D) $208 \ cm^2$

23. Data: $40, 42, 43, 44, 45, 46, 48$. Find the range and describe the spread.

Your Answer

Find more at
ViewMath.com/OK-Grade6

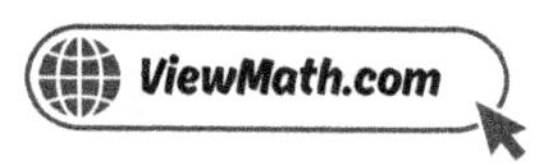

24. The bar graph below shows test scores for 5 students.

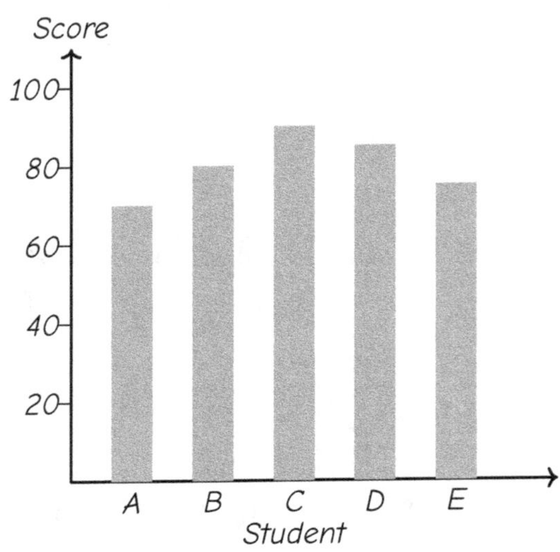


The scores are 70, 80, 90, 85, 75. What is the mean score?

(A) 75

(B) 80

(C) 85

(D) 90

25. Can you find the exact mode from a histogram? Explain.

Your Answer

26. Data: $1, 3, 5, 7, 9, 11, 13, 15, 17, 19$. What is the IQR?

(A) 8

(B) 10

(C) 14

(D) 18

27. Two box plots are shown for heights (in cm) of plants grown with Fertilizer A and Fertilizer B.

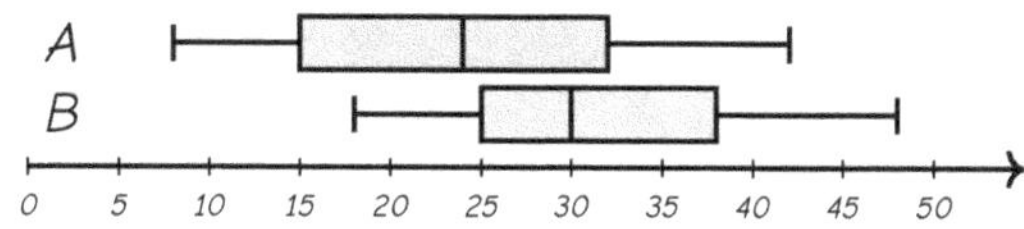

Write the five-number summary for each fertilizer.  Compare the medians, ranges, and IQRs.  Which fertilizer produced taller plants on average? Which produced more consistent growth?

Your Answer:

28. Which of the following best describes the range of all probabilities?

(A) From −1 to 1

(B) From 0 to 1

(C) From 0 to 100

(D) From 1 to 10

29. A student made the stem-and-leaf plot below.  What is wrong with it?

| Stem | Leaves |
| --- | --- |
| 5 | 3 1 7 9 |
| 6 | 2 4 |

Key: 5 | 3 means 53

(A) The stems are out of order.

(B) The leaves for stem 5 are not in order from least to greatest.

(C) The key is incorrect.

(D) There are too few stems.

30. A double bar graph compares boys and girls who participated in a fun run.  The bars show 24 boys and 30 girls. What fraction of all participants were boys?

(A) $\dfrac{24}{30}$

(B) $\dfrac{4}{9}$

(C) $\dfrac{5}{9}$

(D) $\dfrac{1}{2}$

 # End of Practice Test 7 

*Great job finishing the test!*

 **My Score**

I got __________ out of 30 questions right.

*Check your answers in the **Answer Key** at the back of the book.*

*Review any questions you missed. That's how we learn!*

**Check Your Score Online!**

Visit **ViewMath Academy** to enter your answers and see which topics you need to review. You can also explore lessons, take quizzes, track your scores, and save your progress!

viewmath.com/score/6.1.OK.22

*Or go to viewmath.com/score and enter code: 6.1.OK.22*

# Practice Test 8

 30 Questions

## ✏ Before You Start ✏

- ✓ **Read each question carefully** before choosing your answer.
- ✓ **Show your work** on scratch paper when you need to.
- ✓ **Skip hard questions** and come back to them later.
- ✓ **Check your answers** when you're done.
- ✓ **Take your time** — there's no rush!

⭐ **You've Got This!** ⭐

Do your best and show what you know!

1. A faucet drips 24 drops in 4 minutes. Which statement describes this rate?

   (A) The ratio of drops to drops is 24 : 4.

   (B) The faucet drips at a rate of 24 drops per 4 minutes.

   (C) The faucet drips 4 drops per 24 minutes.

   (D) The total is 28.

2. The graph below shows points that represent equivalent ratios of flour to sugar in a recipe.

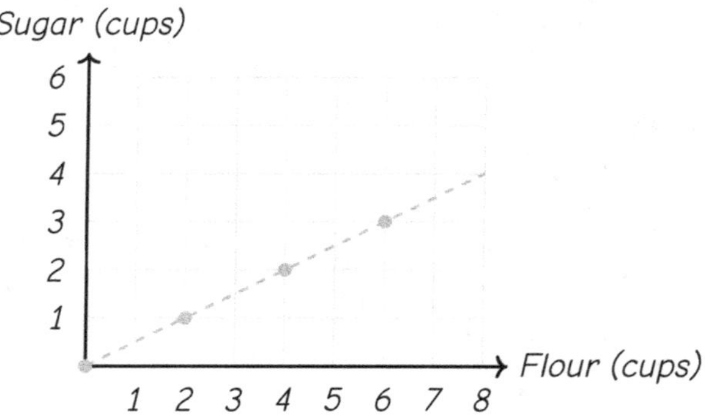


**Part A:** What is the ratio of flour to sugar?

**Part B:** If you need 8 cups of flour, how many cups of sugar do you need?

Your Answer:

3. True or false: a graph of equivalent ratios can be a curved line. Explain.

Your Answer:

4. A class of 30 students was surveyed. 12 chose chocolate ice cream. What percent chose chocolate?

   (A) 12%

   (B) 30%

   (C) 40%

   (D) 60%

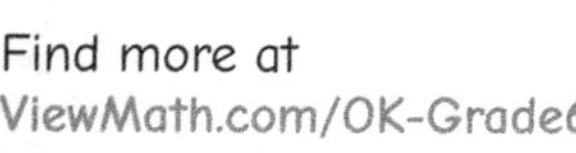

5. A truck can carry 3 tons. A shipment weighs 5,500 pounds. Can the truck carry the shipment? (1 ton = 2,000 pounds)

*Your Answer*

6. A floor plan of an apartment is shown below. The scale is 1 cm = 6 m.

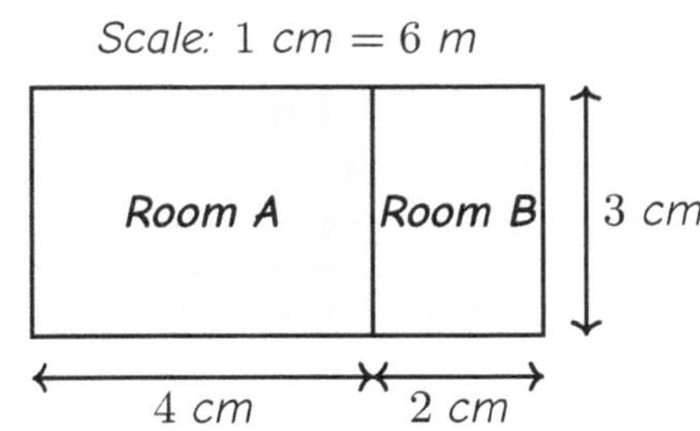


What is the total actual area of Room A and Room B combined?

(A) 108 square meters

(B) 216 square meters

(C) 432 square meters

(D) 648 square meters

7. Kai says $\dfrac{2}{7} \div \dfrac{5}{6} = \dfrac{12}{35}$. Is Kai correct?

(A) No, the answer is $\dfrac{10}{42}$

(B) Yes, $\dfrac{12}{35}$ is correct

(C) No, the answer is $\dfrac{35}{12}$

(D) No, the answer is $\dfrac{7}{15}$

8. What is $7,344 \div 24$?

(A) 36

(B) 360

(C) 306

(D) 3,006

Find more at
ViewMath.com/OK-Grade6

ViewMath.com

9. *What is the opposite of 0?*

   (A)  1                                          (B)  $-1$

   (C)  0                                          (D)  *There is no opposite of* 0

10. **Part A:** *In which quadrant is the point* $(-8, 1)$ *located?*

   **Part B:** *If* $(-8, 1)$ *is reflected across the* $y$-*axis, what are the coordinates of its image and in which quadrant does it lie?*

   **Part C:** *If the original point* $(-8, 1)$ *is instead reflected across the* $x$-*axis, what are the coordinates of its image and in which quadrant does it lie?*

   Your Answer

11. *The table below shows powers of 3. What value belongs in the blank?*

| $3^1$ | $3^2$ | $3^3$ | $3^4$ | $3^5$ |
|-------|-------|-------|-------|-------|
| 3     | 9     | 27    | ?     | 243   |

   (A)  36                                         (B)  64

   (C)  81                                         (D)  108

12. *Which expression represents "the sum of 9 and the product of 4 and* $t$*"?*

   (A)  $9 + 4 + t$                                (B)  $9 + 4t$

   (C)  $9(4 + t)$                                 (D)  $9 \times 4t$

13. A parking garage charges $5 plus $3 per hour. How much does it cost to park for 6 hours? Use the expression $5 + 3h$.

Your Answer:

14. The rectangle below has ...sions shown. Write a simplified expression for its perimeter.

$$2x + 5$$
$$x + 1 \qquad x + 1$$
$$2x + 5$$

...:ball cards. He buys 12 more and then gives 5 to his friend. Write an expression for how ...w.

...ve: $15a = 105$

Your Answer:

17. Is $n = 4$ a solution to $n > 4$? Is $n = 4$ a solution to $n \geq 4$? Explain both.

Your Answer:

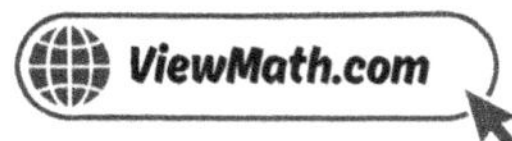

18. Which of these numbers is a solution to $x \geq -3$?

(A) $-4$

(B) $-3.5$

(C) $-3$

(D) $-100$

19. A bathtub drains at 3 gallons per minute. It starts with 45 gallons. Write an e
after $t$ minutes. When is the tub empty?

*the water $w$ remaining*

Your Answer

20. A student says the area of this parallelogram is 48 square units. The base is 8 and
actual height, shown by the dashed line, is 5. Is the student correct?

*The*

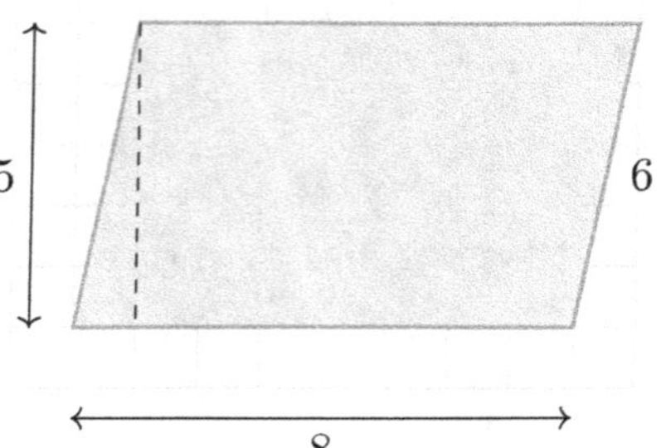

5

6

8

(A) Yes, because $8 \times 6 = 48$.

(B) No, the area is 40 because the height is 5, not 6.

(C) No, the area is 24 because you must divide by 2.

(D) Yes, because $\frac{1}{2} \times 8 \times 6 = 24$ and you double it.

21. A right triangle has vertices $(-3, -2)$, $(5, -2)$, and $(-3, 4)$. What is the area?

Your Answer:

22. A cube has edges of 5 in. What is the surface area?

(A) $25\ in^2$

(B) $125\ in^2$

(C) $150\ in^2$

(D) $75\ in^2$

23. Data set: $22, 24, 25, 25, 26, 27, 45$. Which value is most likely an outlier?

(A) 22

(B) 25

(C) 27

(D) 45

24. The mean of four numbers is 30. Three of the numbers are 25, 28, and 35. What is the fourth number?

Your Answer:

25. A dot plot of daily steps: 3000 (1), 4000 (2), 5000 (6), 6000 (4), 7000 (2). What is the mode? What is the range?

Your Answer:

26. Data (in order): $3, 5, 7, 9, 11, 13, 15$. What is the median?

(A) 7

(B) 9

(C) 11

(D) 13

27. Data: $4, 8, 10, 12, 14, 16, 100$. Find the mean and median. Which measure better represents a typical value? Why?

Your Answer:

Find more at
ViewMath.com/OK-Grade6

28. A spinner is divided into 5 equal sections numbered 1 through 5.  What is the probability of landing on a number greater than 3?

(A) $\dfrac{3}{5}$

(B) $\dfrac{2}{5}$

(C) $\dfrac{1}{5}$

(D) $\dfrac{4}{5}$

29. Find the median of the data in the stem-and-leaf plot below.

| Stem | Leaves |
|---|---|
| 2 | 3 5 8 |
| 3 | 0 2 6 7 |
| 4 | 1 4 |

Key: 2 | 3 means 23

Your Answer

30. In a frequency table, the colors chosen by students are: red (12), blue (7), green (7), yellow (4).  How many more students chose red than yellow?

(A) 4

(B) 5

(C) 7

(D) 8

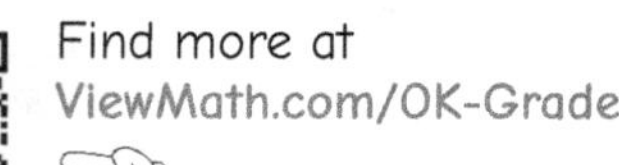

#  End of Practice Test 8 

## Great job finishing the test!

### ☑ My Score

I got ___________ out of 30 questions right.

Check your answers in the **Answer Key** at the back of the book.

💡 Review any questions you missed. That's how we learn!

### 📊 Check Your Score Online!

Visit **ViewMath Academy** to enter your answers and see which topics you need to review. You can also explore lessons, take quizzes, track your scores, and save your progress!

viewmath.com/score/6.1.OK.23

Or go to viewmath.com/score and enter code:  6.1.OK.23

# Practice Test 9

 30 Questions

 Before You Start

- ✓ **Read each question carefully** before choosing your answer.
- ✓ **Show your work** on scratch paper when you need to.
- ✓ **Skip hard questions** and come back to them later.
- ✓ **Check your answers** when you're done.
- ✓ **Take your time** — there's no rush!

 You've Got This! 

Do your best and show what you know!

1. Mia walks 3 kilometers in 45 minutes. Leo walks 5 kilometers in 60 minutes. Who walks faster?

   (A) Mia

   (B) Leo

   (C) They walk at the same speed.

   (D) Cannot be determined.

2. The ratio of red to blue beads is $1 : 4$. If there are 3 red beads, how many blue beads are there?

   (A) 4

   (B) 7

   (C) 12

   (D) 8

3. Two ratio graphs both pass through the origin. Line A goes through $(1, 4)$ and Line B goes through $(1, 3)$. Which line is steeper?

   (A) Line A

   (B) Line B

   (C) They have the same steepness.

   (D) Cannot be determined.

4. A jacket costs $80. It is 25% off. How much do you save?

   (A) $25

   (B) $20

   (C) $60

   (D) $15

5. Convert 3 gallons to pints. $(1 \text{ gallon} = 8 \text{ pints})$

   (A) 16

   (B) 24

   (C) 32

   (D) 11

6. A blueprint has a scale of 1 inch = 6 feet. A room is 2.5 inches wide on the blueprint. What is the actual width of the room?

(A) 8.5 feet

(B) 12 feet

(C) 15 feet

(D) 36 feet

7. Each serving of juice is $\frac{2}{5}$ cup. A jug holds $\frac{4}{5}$ cup. How many servings are in the jug?

(A) $\frac{8}{25}$

(B) $\frac{2}{5}$

(C) $\frac{5}{2}$

(D) 2

8. The long division below for $1{,}575 \div 5$ is partially completed. A digit in the quotient is missing.

$$\begin{array}{r} 3\ \square\ 5 \\ 5\ \overline{\smash{)}\ 1\,5\,7\,5} \\ -1\,5\phantom{\,7\,5} \\ \overline{\phantom{-1}\ 7\phantom{\,5}} \\ -5\phantom{\,5} \\ \overline{\phantom{-}2\,5} \\ -2\,5 \\ \overline{\phantom{-2}0} \end{array}$$

What digit belongs in the box? Explain how you know.

Your Answer.

9. A number plus its opposite always equals which value?

   (A) 1                                    (B) The number itself

   (C) −1                                   (D) 0


10. A triangle is drawn on the coordinate plane below with vertices at A, B, and C.

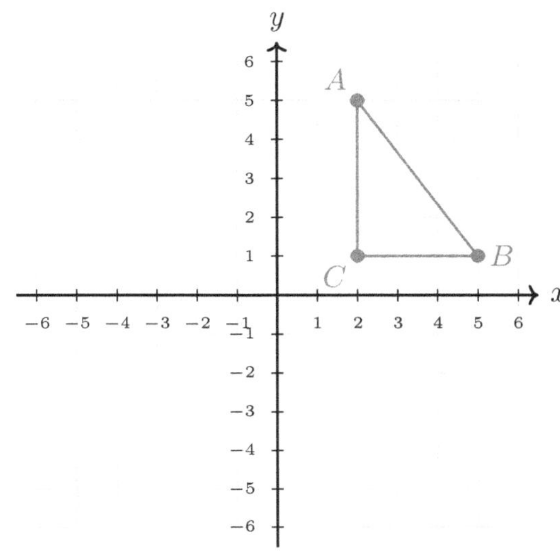

**Part A:** Write the coordinates of each vertex: A, B, and C.

**Part B:** If the triangle is reflected across the y-axis, write the coordinates of the new vertices A', B', and C'.

**Part C:** In which quadrant will the reflected triangle be located?


11. Evaluate: $(5 - 1)^2 + 7$

   (A) 14                                   (B) 23

   (C) 11                                   (D) 33

12. Which expression represents "a number $n$ plus 8"?

  (A) $n - 8$          (B) $8n$

  (C) $n + 8$          (D) $n \div 8$

13. Evaluate $6x - x^2$ when $x = 4$.

  (A) 8          (B) 20

  (C) $-8$          (D) 40

14. Simplify: $7n + 4n$

  (A) $11n$          (B) $28n$

  (C) $11n^2$          (D) $74n$

15. In $P = 4s$, can $s$ be any positive number, or is it one specific number?

  (A) Only $s = 1$          (B) Only whole numbers

  (C) Any positive number — the formula works for          (D) Only $s = 4$
all squares

16. Solve: $\dfrac{t}{6} = 5$

  (A) $t = 11$          (B) $t = 1$

  (C) $t = 30$          (D) $t = 56$

Find more at
ViewMath.com/OK-Grade6

Get Online

**ViewMath.com**

17. Is $x = 5$ a solution to $x > 5$?

(A) Yes, because 5 is equal to 5

(B) Yes, because 5 is positive

(C) No, because 5 is not greater than 5

(D) No, because 5 is greater than 5

18. Which inequality matches a graph with an open circle at $-2$ and shading to the left?

(A) $x > -2$

(B) $x < -2$

(C) $x \geq -2$

(D) $x \leq -2$

19. In $c = 7p$, where $c$ is total cost and $p$ is the number of pizzas, what is the cost of 6 pizzas?

(A) \$13

(B) \$36

(C) \$42

(D) \$67

20. A field is shapea like a parallelogram with base 25 m and height 14 m. Each bag of seed covers 50 $m^2$. How many bags are needed?

(A) 5 bags

(B) 6 bags

(C) 7 bags

(D) 8 bags

21. A right triangle has vertices $(1, 1)$, $(1, 9)$, and $(5, 1)$. What is the area?

(A) 32 square units

(B) 16 square units

(C) 20 square units

(D) 8 square units

Find more at
ViewMath.com/OK-Grade6

**ViewMath.com**

22. *What is the surface area of a rectangular prism with length 6 cm, width 4 cm, and height 3 cm?*

   (A) *72 cm²*             (B) *108 cm²*

   (C) *54 cm²*             (D) *96 cm²*

23. *Data: 4, 5, 5, 6, 6, 6, 7, 7, 8. Which feature best describes this data?*

   (A) *It has a large gap.*             (B) *It has an outlier at 4.*

   (C) *It clusters around 6 with a peak there.*        (D) *It is extremely spread out.*

24. *What is the median of the data set 3, 7, 9, 15, 21?*

   (A) 7             (B) 9

   (C) 11             (D) 15

25. *Look at the frequency table below.*

| Favorite Sport | Frequency |
| --- | --- |
| Soccer | 9 |
| Basketball | 7 |
| Baseball | 5 |
| Swimming | 4 |

*What fraction of students chose Basketball?*

   (A) $\frac{7}{9}$             (B) $\frac{7}{25}$

   (C) $\frac{7}{16}$             (D) $\frac{7}{12}$

Find more at
ViewMath.com/OK-Grade6

26. If the whiskers of a box plot are both short and the box is narrow, what does this tell you?

(A) The data is very spread out.

(B) The data is highly consistent with little variability.

(C) There are many outliers.

(D) The data set is very large.

27. A dot plot shows test scores for Class 1: most dots clustered between 75 and 85 with one dot at 40. Class 2's dot plot shows dots evenly spread from 55 to 95. Which class is likely more consistent?

(A) Class 1, because most scores are close together.

(B) Class 2, because the scores are evenly spread.

(C) Neither — they are equally consistent.

(D) Cannot be determined without the means.

28. A standard number cube is rolled. What is the probability of rolling an even number?

(A) $\dfrac{1}{6}$

(B) $\dfrac{1}{3}$

(C) $\dfrac{1}{2}$

(D) $\dfrac{2}{3}$

29. A key on a stem-and-leaf plot reads $5 \mid 3 = 53$. What value does $5 \mid 0$ represent?

(A) 5

(B) 50

(C) 53

(D) 500

30. A dcuble bar graph shows the number of apples and oranges sold each day. On Monday, 15 apples and 20 oranges were sold. On Tuesday, 25 apples and 10 oranges were sold. On which day were more total fruits sold, and how many more?

Your Answer

Find more at
ViewMath.com/OK-Grade6

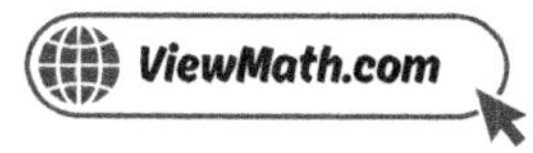

#  End of Practice Test 9 

Great job finishing the test!

##  My Score

I got _____________ out of 30 questions right.

*Check your answers in the **Answer Key** at the back of the book.*

💡 *Review any questions you missed. That's how we learn!*

## 📊 Check Your Score Online!

Visit **ViewMath Academy** to enter your answers and see which topics you need to review. You can also explore lessons, take quizzes, track your scores, and save your progress!

viewmath.com/score/6.1.OK.24

Or go to viewmath.com/score and enter code: 6.1.OK.24

# Practice Test 10

 30 Questions

## ✏️ Before You Start ✏️

- ✓ **Read each question carefully** before choosing your answer.
- ✓ **Show your work** on scratch paper when you need to.
- ✓ **Skip hard questions** and come back to them later.
- ✓ **Check your answers** when you're done.
- ✓ **Take your time** — there's no rush!

 You've Got This! 

Do your best and show what you know!

1.  The table below shows the distance a car traveled over time.

| Hours | Miles |
|-------|-------|
| 1 | 55 |
| 2 | 110 |
| 3 | 165 |
| 4 | 220 |

What is the rate in miles per hour?

(A) 110 miles per hour

(B) 55 miles per hour

(C) 220 miles per hour

(D) 4 miles per hour

2.  The ratio table below shows the relationship between bags of seed and the area of garden they cover.

| Bags of Seed | Area (sq ft) |
|--------------|--------------|
| 2 | 50 |
| 4 | 100 |
| 6 | ? |
| 8 | 200 |

What is the missing value?

(A) 125

(B) 150

(C) 175

(D) 160

3.  Juice costs \$2 per bottle. Which point would **not** appear on the graph of this ratio?

(A) $(1, 2)$

(B) $(3, 6)$

(C) $(4, 10)$

(D) $(5, 10)$

Find more at
ViewMath.com/OK-Grade6

**ViewMath.com**

4. Maria saved \$45 toward a \$90 bike. What percent has she saved?

(A) 25%

(B) 45%

(C) 50%

(D) 75%

5. Which unit conversion uses a ratio?

(A) $5 + 12 = 17$ inches

(B) $\dfrac{5\ ft}{1} \times \dfrac{12\ in}{1\ ft} = 60\ in$

(C) $5 - 12 = -7$ inches

(D) $\dfrac{12}{5} = 2.4$ feet

6. A map shows three towns connected by roads. The scale is 1 cm = 15 km.

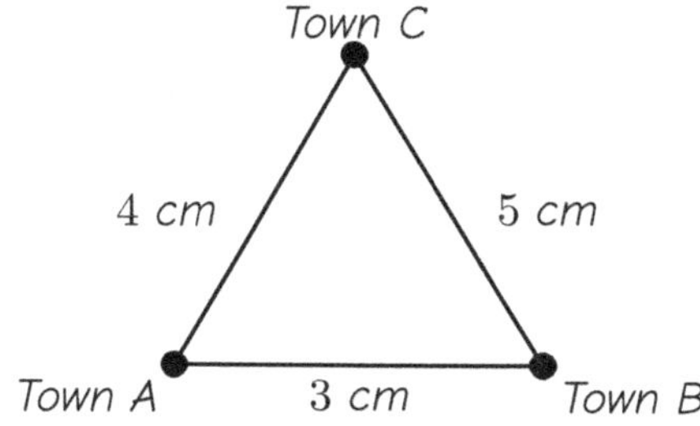


**Part A:** Find the actual distance from Town A to Town B.

**Part B:** Find the actual distance from Town B to Town C.

**Part C:** A delivery truck travels the complete route A → B → C → A. What is the total actual distance of the round trip?

Your Answer:

7. Evaluate: $\dfrac{7}{10} \div \dfrac{2}{5}$

Your Answer:

8. *Compute* $6{,}048 \div 21$. *Then check your answer using multiplication.*

*Your Answer:*

9. *What is* $|0|$*?*

(A) *There is no answer*

(B) $-1$

(C) $1$

(D) $0$

10. *In which quadrant is the point* $(2, -5)$ *located?*

(A) *Quadrant I*

(B) *Quadrant II*

(C) *Quadrant III*

(D) *Quadrant IV*

11. *The area of each small square below is* 1 *square unit. Write the area using an exponent, then find the total area.*

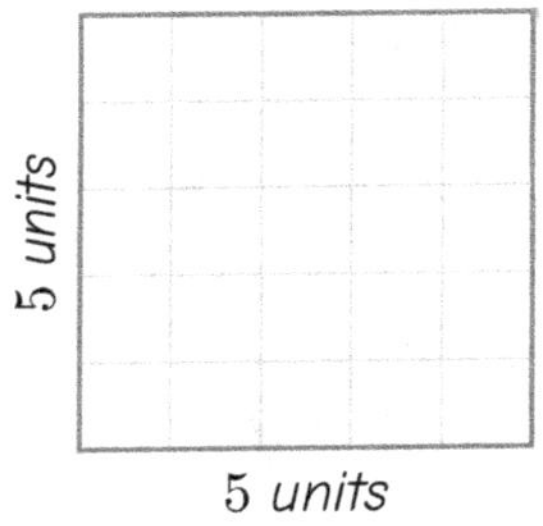

5 units (vertical) — 5 units (horizontal)

*Your Answer:*

12. *What does the expression* $h + 12$ *represent if* $h$ *is Hannah's height in inches?*

(A) *Hannah's height minus 12 inches*

(B) *A height that is 12 inches more than Hannah's*

(C) *Hannah's height times 12*

(D) *12 inches divided by Hannah's height*

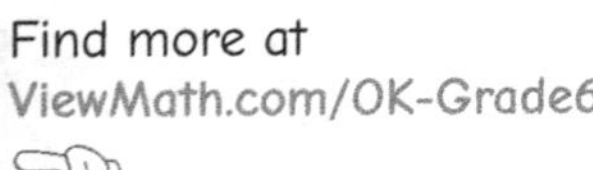
Find more at
ViewMath.com/OK-Grade6

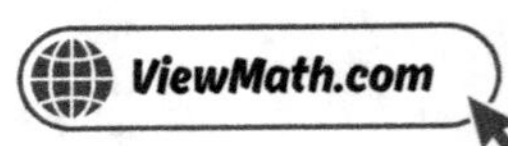

13. The table below shows the cost of renting a bike using the expression $C = 8 + 3h$, where $h$ is the number of hours. Fill in the missing values A and B.

| Hours ($h$) | Cost ($C$) |
| --- | --- |
| 1 | $11 |
| 2 | A |
| 3 | $17 |
| 5 | B |

Your Answer

14. Which expression is equivalent to $7(k - 3) + 5$?

(A) $7k - 16$

(B) $7k + 2$

(C) $7k - 26$

(D) $7k - 21$

15. You have 100 pages to read and read $r$ pages each day. Which expression tells how many pages are left after 4 days?

(A) $100 + 4r$

(B) $100 - 4r$

(C) $4r - 100$

(D) $100r - 4$

16. Solve: $m + 17 = 30$

Your Answer



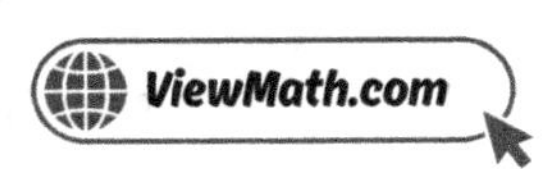

17. Which inequality represents "you need more than \$25 to buy the video game"?

  (A) $d \le 25$            (B) $d < 25$

  (C) $d > 25$            (D) $d \ge 25$

18. Look at the number line below and write the inequality it represents.

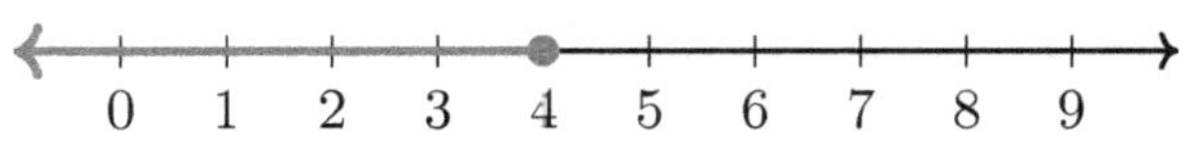

Your Answer:

19. Which equation represents this relationship: "The number of legs $L$ is 4 times the number of dogs $d$"?

  (A) $d = 4L$            (B) $L = d + 4$

  (C) $L = 4d$            (D) $L = d \div 4$

20. A trapezoid has bases $b_1 = 5$ ft and $b_2 = 13$ ft, and height $h = 8$ ft. What is the area?

  (A) $144 \text{ ft}^2$            (B) $72 \text{ ft}^2$

  (C) $52 \text{ ft}^2$            (D) $36 \text{ ft}^2$

21. A rectangle has vertices $(0,0)$, $(6,0)$, $(6,4)$, and $(0,4)$. What is the area?

  (A) 10 square units            (B) 20 square units

  (C) 24 square units            (D) 12 square units

Find more at
ViewMath.com/OK-Grade6

ViewMath.com

22. Which pair of dimensions gives a rectangular prism with the greatest surface area?

(A) $2 \times 3 \times 10$

(B) $4 \times 4 \times 4$

(C) $1 \times 5 \times 8$

(D) $3 \times 3 \times 6$

23. Data: $60, 62, 64, 66, 68, 70$. Is this data symmetric? Explain.

Your Answer:

24. A frequency table shows how many books students read last month:

| Books | Students |
|-------|----------|
| 1 | 3 |
| 2 | 5 |
| 3 | 4 |
| 4 | 2 |
| 5 | 1 |

What is the median number of books?

(A) 2

(B) 2.5

(C) 3

(D) 4

25. You want to know whether to display data as a dot plot, histogram, or frequency table. Which question should you ask first?

(A) What is the mean of the data?

(B) How many data points are there, and is the data numerical or categorical?

(C) What color should the bars be?

(D) Is the data symmetric or skewed?

26.  *Data (in order):* $5, 5, 6, 7, 8, 9, 10, 10, 12, 15, 18, 20.$ *Find Q1 and Q3.*



27.  *Class A test scores: median = 80, range = 30, IQR = 10. Class B test scores: median = 75, range = 45, IQR = 22. Which class performed better overall and which was more consistent?*

(A) *Class B performed better and was more consistent.*

(B) *Class A performed better and was more consistent.*

(C) *Class A performed better, Class B was more consistent.*

(D) *Class B performed better, Class A was more consistent.*

28.  *A spinner is divided into 6 equal sections as shown. What is the probability of spinning* **blue**?

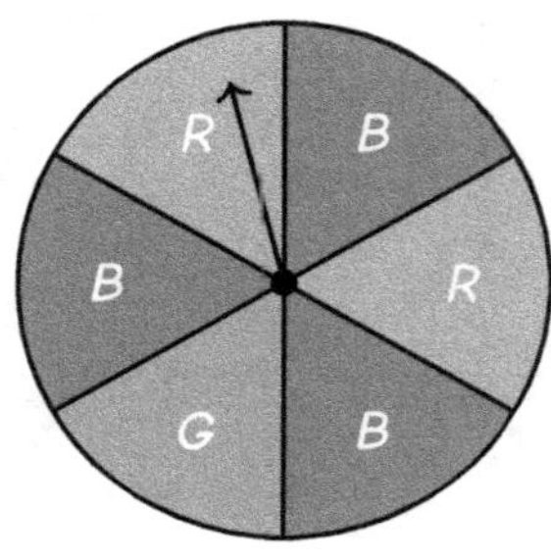


(A) $\dfrac{1}{6}$

(B) $\dfrac{1}{3}$

(C) $\dfrac{1}{2}$

(D) $\dfrac{2}{3}$

29.  *A stem-and-leaf plot uses stems* $10, 11, 12$ *with a key* $10 \mid 2 = 102.$ *What data value does stem* 11 *with leaf* 5 *represent?*

(A) 15

(B) 115

(C) 150

(D) 1105

30. A frequency table shows test scores. The title of the table is "Math Quiz Scores for Period 3." What does the title tell you?

(A) The scores are from a science test

(B) The scores are from every class in the school

(C) The scores are math quiz results from one specific class period

(D) The table shows grades for the whole year



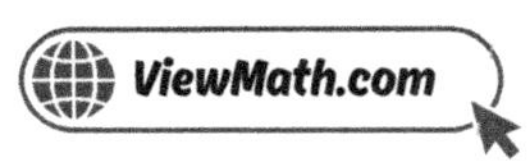

 # ★ End of Practice Test 10 ★ 

Great job finishing the test!

 **My Score**

I got ____________ out of 30 questions right.

*Check your answers in the **Answer Key** at the back of the book.*

💡 *Review any questions you missed. That's how we learn!*

📊 **Check Your Score Online!**

Visit **ViewMath Academy** to enter your answers and see which topics you need to review. You can also explore lessons, take quizzes, track your scores, and save your progress!

viewmath.com/score/6.1.OK.25

*Or go to viewmath.com/score and enter code: 6.1.OK.25*

# Answer Key & Explanations

# Answer Key

First try each test on your own, then check your work here.

## ☑ Practice Test 1 — Answer Key

**1** Answers will vary. Example: "Apples cost $2 per pound."    **2** C    **3** B    **4** A    **5** C

**6** The student divided $6 \div 4 = 1.5$ instead of multiplying. The correct actual length is $6 \times 4 = 24$ feet.    **7** A

**8** D    **9** 15    **10** $(-7, -3)$    **11** A    **12** $3n - 8$    **13** 22    **14** B

**15** Answers vary. Example: You buy $n$ notebooks at $5 each and pay $3 for shipping.    **16** C

**17** Answers vary. Example: A classroom has at most 15 computers.

**18** $x = 3$ IS a solution to $x \le 3$ because $3 \le 3$ is true. $x = 3$ is NOT a solution to $x < 3$ because $3 < 3$ is false.

**19** B    **20** A    **21** B    **22** A    **23** B    **24** B    **25** C    **26** B    **27** B    **28** 20%

**29** B    **30** positive trend

## 💡 Time to Learn! 💡

Review the explanations below, **especially for the questions you missed**.

Understanding why each answer is correct builds stronger problem-solving skills.

**Tip:** Circle any questions you got wrong, then read their explanation carefully.

Find more at
ViewMath.com/OK-Grade6

VIewMath.com

## 📖 Practice Test 1 — Detailed Explanations

**1**   Any sentence that compares a dollar amount to a weight in pounds is correct. For example, "Ground beef costs \$5 per 2 pounds."

**2**   From the graph, each hour earns \$10 (2 units on the y-axis = \$10). In 6 hours: $6 \times 10 = \$60$.

**3**   Ratio: $1 : 4$. When $x = 3$: $y = 3 \times 4 = 12$.

**4**   $0.15 \times 60 = 9$.

**5**   1 yard = 3 feet. $7 \times 3 = 21$ feet.

**6**   To find the actual length from a drawing, multiply the drawing length by the scale factor. $6 \times 4 = 24$ feet.

**7**   The number line shows 5 jumps of $\frac{1}{6}$ to reach $\frac{5}{6}$. This models $\frac{5}{6} \div \frac{1}{6} = 5$.

**8**   $1{,}344 \div 8 = 168$ pages per week. Check: $168 \times 8 = 1{,}344$.

**9**   $|-9| = 9$ and $|-6| = 6$. Adding these gives $9 + 6 = 15$. Find each absolute value first, then add.

**10**   Reflecting across the $x$-axis changes the sign of the $y$-coordinate: $(x, y) \rightarrow (x, -y)$. So $(-7, 3)$ becomes $(-7, -3)$.

**11**   Test $n = 1$: $1^2 + 2 = 3$ ✓. $n = 2$: $4 + 2 = 6$ ✓. $n = 3$: $9 + 2 = 11$ ✓. $n = 4$: $16 + 2 = 18$ ✓.

**12**   Start with $n$. Multiply by 3: $3n$. Subtract 8: $3n - 8$.

**13** $3(8 + 2) - 8 = 3(10) - 8 = 30 - 8 = 22.$

**14** *Distribute:* $3 \times x + 3 \times 5 = 3x + 15.$

**15** $5n$ *represents a cost per item and 3 represents a fixed fee.*

**16** *Multiply both sides by 4:* $m = 7 \times 4 = 28.$

**17** *Any situation where a quantity is 15 or fewer works.*

**18** $\leq$ *includes the boundary value;* $<$ *does not.*

**19** *In B,* $y$ *is always 7 no matter what* $x$ *is. The value of* $y$ *does not change with* $x$.

**20** *Rectangle:* $8 \times 5 = 40.$ *Triangle:* $\frac{1}{2} \times 8 \times 3 = 12.$ *Total:* $40 + 12 = 52$ $cm^2$.

**21** *Base* $= |2 - (-4)| = 6.$ *Height* $= |5 - 0| = 5.$ *Area* $= \frac{1}{2} \times 6 \times 5 = 15$ *square units.*

**22** $SA = 2(12)(8) + 2(12)(5) + 2(8)(5) = 192 + 120 + 80 = 392$ $in^2$.

**23** *Symmetric data looks roughly even on both sides of the center when displayed on a graph.*

**24** *Median (middle value) is still 35 since it's the 3rd value. Mean increases from 35 to* $(25+30+35+40+100) \div 5 = 46.$

**25** *Size 8 appears 3 times (6: 1, 7: 2, 8: 3, 9: 2, 10: 1).*

**26** *A long right whisker means some data values are much higher than Q3. This indicates right skew.*

Find more at
ViewMath.com/OK-Grade6

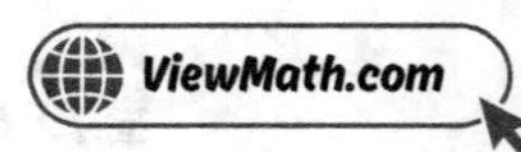

27  Restaurant B's median (10) is lower ⇒ shorter typical wait. Restaurant A's IQR (4) is smaller ⇒ more predictable waits.

28  Total marbles: $4 + 6 + 10 = 20$. $P(green) = \dfrac{4}{20} = \dfrac{1}{5} = 0.20 = 20\%$.

29  Values greater than 50: $51, 53, 56, 62$. That is 4 values.

30  Since the temperature increases each month from January to May, the line goes upward, which is a positive trend.

## ✅ Practice Test 2 — Answer Key

1  A     2  12, 16, and 30     3  Part A: Runner A $= 3 : 4$; Runner B $= 2 : 3$.     Part B: Runner A is faster.

4  C     5  C     6  C     7  B     8  301     9  B

10  The point $(9, 0)$ lies on the $x$-axis because its $y$-coordinate is 0. Points on the axes are not inside any quadrant.

11  B     12  $\dfrac{t}{4} - 1$     13  30 square cm     14  $6a + 10$     15  A     16  B     17  C     18  C

19  Independent: $h$ (hours). Dependent: $d$ (distance).     20  B     21  A     22  268 $m^2$     23  B

24  Mean $= 12$, Median $= 12$     25  B     26  Min $= 2$, $Q1 = 4.5$, Median $= 8$, $Q3 = 11$, Max $= 15$. IQR $= 6.5$.

27  B     28  C     29  B     30  B

Find more at
ViewMath.com/OK-Grade6

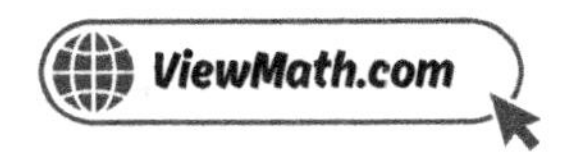

💡 *Time to Learn!* 💡

*Review the explanations below, **especially for the questions you missed**.*

*Understanding why each answer is correct builds stronger problem-solving skills.*

***Tip:*** *Circle any questions you got wrong, then read their explanation carefully.*

## 📖 Practice Test 2 — Detailed Explanations

**1**  *Rosa: 4 cans in 2 days = 2 cans per day. Jake: 3 cans in 2 days = 1.5 cans per day. Rosa has the higher rate.*

**2**  *Row 2: $4 \times 2 = 8$, so $6 \times 2 = 12$. Row 3: $6 \times 4 = 24$, so $4 \times 4 = 16$. Row 4: $4 \times 5 = 20$, so $6 \times 5 = 30$.*

**3**  *Runner A: 3 laps in 4 min = 0.75 laps/min. Runner B: 2 laps in 3 min $\approx$ 0.667 laps/min. $0.75 > 0.667$, so Runner A is faster. On the graph, Runner A's line is steeper.*

**4**  *$0.25 \times ? = 20$. Divide: $20 \div 0.25 = 80$.*

**5**  *$500 \div 1{,}000 = 0.5$ kg.*

**6**  *Actual dimensions: $4 \times 3 = 12$ ft by $5 \times 3 = 15$ ft. Area $= 12 \times 15 = 180$ sq ft. Choice A (20) is just the drawing area.*

**7**  *$\dfrac{3}{5} \times \dfrac{5}{1} = \dfrac{15}{5} = 3$.*

**8**  *$45 \div 15 = 3$; bring down $1 \rightarrow 1 \div 15 = 0$ R1; bring down $5 \rightarrow 15 \div 15 = 1$. Answer: 301. The zero in the tens place is important. Check: $301 \times 15 = 4{,}515$.*

Find more at
ViewMath.com/OK-Grade6

**9** Absolute value tells you how far a number is from zero. $-12$ is 12 units from zero, so $|-12| = 12$. A common mistake is to think $|-12| = -12$, but distance is never negative.

**10** The four quadrants are the regions between the axes. A point is on an axis when $x = 0$ (on the $y$-axis) or $y = 0$ (on the $x$-axis). Since $(9, 0)$ has $y = 0$, it sits on the $x$-axis itself, which is the boundary between Quadrant I and Quadrant IV—not in either one.

**11** Parentheses: $2 + 4 = 6$. Exponent: $6^2 = 36$. Multiply: $3 \times 36 = 108$.

**12** Divide $t$ by 4 to get $\frac{t}{4}$, then subtract 1: $\frac{t}{4} - 1$.

**13** $A = \frac{1}{2} \times 10 \times 6 = \frac{1}{2} \times 60 = 30$ square cm.

**14** $3a + 5a - 2a = 6a$ and $7 + 3 = 10$. Result: $6a + 10$.

**15** Tickets: $9p$. One bucket of popcorn: 6. Total: $9p + 6$.

**16** Subtract 15: $k = 32 - 15 = 17$.

**17** "Fewer than 20" means less than 20: $y < 20$.

**18** Open circle at 0 means 0 is not included. Shading right means greater than: $x > 0$.

**19** You choose the hours (independent). Distance depends on hours: $d = 6h$.

**20** Area uses the height, not the slant side. $A = 12 \times 6 = 72\ m^2$.

**21** Length $= |5 - (-3)| = 9$. Width $= 54 \div 9 = 6$ units.

Find more at
ViewMath.com/OK-Grade6

**22**  $SA = 2(10)(8) + 2(10)(3) + 2(8)(3) = 160 + 60 + 48 = 268 \; m^2.$

**23**  Most scores cluster around 70–76. The center is about 73–75. The 98 is an outlier that would pull the mean higher, but the typical values are in the low-to-mid 70s.

**24**  Sum $= 84.$ Mean $= 84 \div 7 = 12.$ Median (4th of 7) $= 12.$

**25**  The 25–49 bar has height 8, the tallest bar. It is the modal interval.

**26**  9 values, already ordered. Median is the 5th value: 8. Lower half: $2, 4, 5, 7.$ $Q1 = (4 + 5) \div 2 = 4.5.$ Upper half: $9, 10, 12, 15.$ $Q3 = (10 + 12) \div 2 = 11.$ $IQR = 11 - 4.5 = 6.5.$

**27**  The median describes a typical value (center) and the IQR describes how spread out the middle 50% of values are (spread). Together they summarize the data well.

**28**  $60\% = 0.6,$ which is greater than 0.5, so the event is likely.

**29**  Class A scores: $62, 65, 68, 71, 74, 77, 80, 86$ (8 values). Median $= (71 + 74) \div 2 = 72.5.$ Class B scores: $63, 66, 70, 72, 75, 79, 81, 84, 92$ (9 values). Median $= 75$ (the 5th value). Class B has the higher median.

**30**  Boys in Class A: bar reaches 6 units $= 30$ books. Girls in Class A: bar reaches 4 units $= 20$ books. Difference: $30 - 20 = 10.$

## ☑ Practice Test 3 — Answer Key

 B   C   B   B   C   B   A   C   C   B

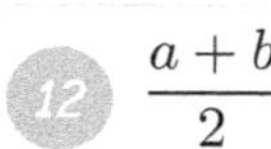 D   $\dfrac{a+b}{2}$  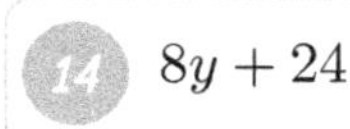 C   $8y + 24$   B   A   D   B   B

Find more at
ViewMath.com/OK-Grade6

 **20** D    **21** A   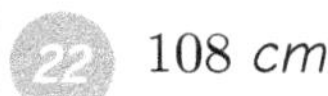 **22** $108\ cm^2$    **23** B    **24** D    **25** C    **26** 50%   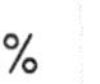 **27** B    **28** C

 **29** B    **30** B

---

## 💡 Time to Learn! 💡

Review the explanations below, **especially for the questions you missed**.

Understanding why each answer is correct builds stronger problem-solving skills.

**Tip:** Circle any questions you got wrong, then read their explanation carefully.

---

## 📖 Practice Test 3 — Detailed Explanations

**1.** People to cups of rice: 6 people per 2 cups. The first quantity mentioned goes first.

**2.** $6 \times 3 = 18$ cups of flour, so $4 \times 3 = 12$ eggs.

**3.** The ratio is $6 : 2 = 3 : 1$. When $x = 15$: $y = 15 \div 3 = 5$.

**4.** Increase: $0.10 \times \$25 = \$2.50$. New price: $\$25 + \$2.50 = \$27.50$.

**5.** $156 \div 12 = 13$ feet.

**6.** Actual distance $= 7 \times 3 = 21$ km.

**7.** $\dfrac{1}{3} \times \dfrac{3}{2} = \dfrac{3}{6} = \dfrac{1}{2}$.

**8.** $15 \div 5 = 3$; bring down $7 \rightarrow 7 \div 5 = 1$ R2; bring down $5 \rightarrow 25 \div 5 = 5$. Answer: 315. Check: $315 \times 5 = 1{,}575$.

Find more at
ViewMath.com/OK-Grade6

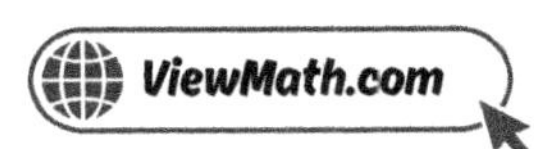

9.  $|6| = 6$ and $|-6| = 6$. Both 6 and $-6$ are 6 units from zero, so both have an absolute value of 6.

10. In Quadrant II, the $x$-coordinate is negative (left of the $y$-axis) and the $y$-coordinate is positive (above the $x$-axis), giving the sign convention $(-, +)$.

11. The base 5 is multiplied 3 times: $5^3$.

12. First find the sum $a + b$, then divide by 2: $\frac{a+b}{2}$.

13. $3^2 + 3 = 9 + 3 = 12$.

14. $8 \times y + 8 \times 3 = 8y + 24$.

15. $d = 60t$ means distance equals 60 miles per hour times the number of hours.

16. Subtract 24: $x = 50 - 24 = 26$.

17. "At most" means the value can equal the number or be less, so it should be $\leq$, not $<$.

18. "More than 75" is $t > 75$: open circle (not including 75), shade right (greater values).

19. 2 cups per batch, $b$ batches: $f = 2b$.

20. $A = \frac{1}{2}(40 + 60)(30) = \frac{1}{2}(100)(30) = 1{,}500 \ m^2$.

21. Rectangle area: $7 \times 3 = 21$. Triangle area: $\frac{1}{2} \times 2 \times 3 = 3$. Remaining: $21 - 3 = 18$ square units.

Find more at
ViewMath.com/OK-Grade6

22. 2 triangles: $2 \times \frac{1}{2}(3)(4) = 12$ cm$^2$. 3 rectangles: $8 \times 3 + 8 \times 4 + 8 \times 5 = 24 + 32 + 40 = 96$ cm$^2$. Total: $12 + 96 = 108$ cm$^2$.

23. Data set A range: $24 - 20 = 4$. Data set B range: $34 - 10 = 24$. Data set B is much more spread out.

24. Mean = sum $\div 5 = 20$, so sum = $20 \times 5 = 100$.

25. Every value is 5, so 5 is the value that appears most often. The mode is 5.

26. Q1 to Q3 is the box, which always contains 50% of the data.

27. X: $Q_1 = 20$, $Q_3 = 60$, $IQR = 40$. Y: $Q_1 = 37$, $Q_3 = 43$, $IQR = 6$. Y is much smaller.

28. A fair coin has 2 equally likely outcomes (heads or tails). $P(\text{heads}) = \frac{1}{2}$.

29. The smallest value is 23 and the largest is 52. Range $= 52 - 23 = 29$.

30. "At least 2" means 2 or more. Students with 2 pets: 5; with 3 pets: 4. Total: $5 + 4 = 9$.

## ✅ Practice Test 4 — Answer Key

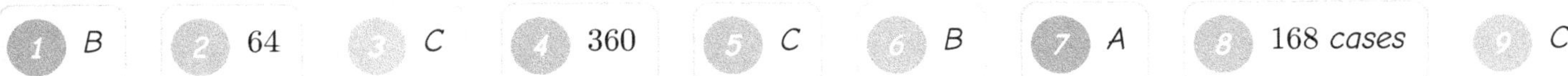

| 1 | B | 2 | 64 | 3 | C | 4 | 360 | 5 | C | 6 | B | 7 | A | 8 | 168 cases | 9 | C |

| 10 | B | 11 | A | 12 | $3n + 2$ | 13 | B | 14 | $11m + 2$ | 15 | B | 16 | A | 17 | D | 18 | B |

19. C  20. C  21. B  22. C  23. D  24. Mean $\approx 13.3$, Median $= 6$. The median is better.

25. B  26. C  27. C  28. B  29. 19  30. B

Find more at
ViewMath.com/OK-Grade6

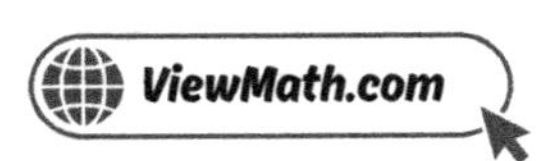

## 💡 Time to Learn! 💡

Review the explanations below, **especially for the questions you missed**.

Understanding why each answer is correct builds stronger problem-solving skills.

**Tip:** Circle any questions you got wrong, then read their explanation carefully.

## 📖 Practice Test 4 — Detailed Explanations

**1** A rate compares two quantities with **different units**. "60 miles per hour" compares miles and hours. The others compare same-type items.

**2** $3 \times 8 = 24$ red, so white $= 5 \times 8 = 40$. Total $= 24 + 40 = 64$.

**3** $5 : 15$ simplifies to $1 : 3$.

**4** $0.72 \times 500 = 360$.

**5** 1 foot $= 12$ inches. $5 \times 12 = 60$ inches.

**6** Drawing length $= 20 \div 4 = 5$ inches. Choice D (80) comes from multiplying instead of dividing.

**7** Dividing by $\dfrac{c}{d}$ is the same as multiplying by its reciprocal, $\dfrac{d}{c}$.

**8** $5{,}376 \div 32 = 168$. Check: $168 \times 32 = 5{,}376$.

**9** Distance from sea level is measured by absolute value. $|-40| = 40$ and $|25| = 25$. Since $40 > 25$, the diver is farther from sea level. The sign tells direction (above or below), not distance.

Find more at
ViewMath.com/OK-Grade6

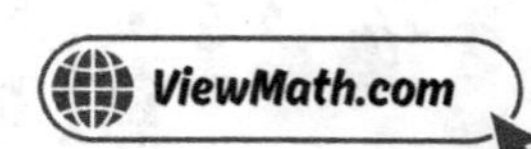

**10** Quadrant II has sign convention $(-,+)$. Since $-4 < 0$ and $7 > 0$, the point $(-4, 7)$ is in Quadrant II. A common mistake is choosing Quadrant III, which has both coordinates negative.

**11** Multiply: $4 \times 3 = 12$. Then left to right: $20 - 12 + 1 = 9$.

**12** $n$ notebooks cost $3n$ dollars. Add the pen: $3n + 2$.

**13** $P = 2(7) + 2(3) = 14 + 6 = 20$ cm.

**14** Distribute: $8m + 2 + 3m$. Combine: $8m + 3m = 11m$. Result: $11m + 2$.

**15** A square has 4 equal sides, so its perimeter is 4 times the side length: $P = 4s$.

**16** Start at unknown $x$, add 8, arrive at 14: $x + 8 = 14$, so $x = 6$.

**17** $2.5 \geq 3$ is false. $2.5 < 3$, so it is not a solution.

**18** $\leq$ means 5 is included (closed circle). Less than 5 is to the left (shade left).

**19** $i = 12(5) = 60$ inches.

**20** The height of a parallelogram is the perpendicular distance from the base to the opposite side, not the slanted side.

**21** Base $= |8 - 2| = 6$. Height $= |7 - 1| = 6$. Area $= \frac{1}{2} \times 6 \times 6 = 18$ square units.

**22** 2 triangular bases $+3$ rectangular faces $= 5$ faces total.

Find more at
ViewMath.com/OK-Grade6

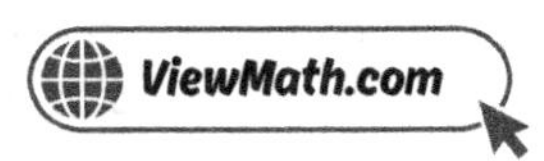

23  Most values are between 2 and 7, clustered around 4. The value 20 is far away from the rest, making it an outlier.

24  Mean: $(3 + 4 + 5 + 6 + 7 + 8 + 60) \div 7 = 93 \div 7 \approx 13.3$. Median: 6. The outlier 60 inflates the mean. The median (6) is more typical.

25  A frequency table lists each value or category and the number of times it appears (its frequency).

26  From Q1 to max covers $50\% + 25\% = 75\%$ of the data (Q1 to Q3 is 50%, Q3 to max is 25%).

27  One extreme outlier pulls the mean away from where most of the data lies. The median would be a better choice.

28  $P(not\ rain) = 1 - P(rain) = 1 - 0.3 = 0.7$.

29  Group X ages (reading leaves from the stem outward): $22, 25, 29, 33, 36, 38, 41$. Range $= 41 - 22 = 19$.

30  A steep upward line means a sharp increase. A flat line means no change. So the value increased sharply then stayed the same.

## 📋 Practice Test 5 — Answer Key

| | | | |
|---|---|---|---|
| 1  50 gallons per hour | 2  B | 3  B | 4  D |
| 5  C | 6  187.5 square feet | 7  C | |
| 8  B | 9  B | 10  (3, −2) | 11  39 |
| 12  C | 13  25 | 14  B | 15  B |
| 16  $w = 72$ | 17  C | 18  C | 19  C |
| 20  B | 21  64 square units | 22  B | 23  D |
| 24  B | 25  B | 26  D | 27  Not always. |
| 28  C | 29  25 | 30  B | |

Get Online

Find more at
ViewMath.com/OK-Grade6

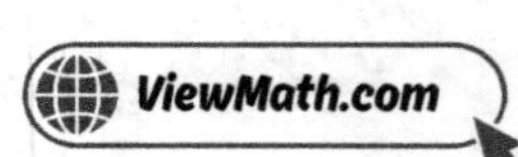

## 💡 Time to Learn! 💡

*Review the explanations below, **especially for the questions you missed**.*

*Understanding why each answer is correct builds stronger problem-solving skills.*

***Tip:** Circle any questions you got wrong, then read their explanation carefully.*

## 📖 Practice Test 5 — Detailed Explanations

**1.** $300 \div 6 = 50$ gallons per hour.

**2.** $2 : 3$ multiplied by 3 gives $6 : 9$. The other pairs do not simplify to $2 : 3$.

**3.** Ratio $= 3 : 2$. Double: $(6, 4)$. Triple: $(9, 6)$, not $(9, 4)$ or $(9, 8)$.

**4.** Backpack: $0.15 \times \$40 = \$6$. Shoes: $0.10 \times \$60 = \$6$. Hat: $0.25 \times \$20 = \$5$. Backpack and shoes are tied at $6.

**5.** $1\ km = 1,000\ m$. $4.5 \times 1,000 = 4,500\ m$.

**6.** Actual dimensions: $3 \times 5 = 15$ ft by $2.5 \times 5 = 12.5$ ft. Area $= 15 \times 12.5 = 187.5$ sq ft.

**7.** $\dfrac{2}{3} \times \dfrac{5}{4} = \dfrac{10}{12} = \dfrac{5}{6}$.

**8.** The four steps of the standard algorithm are **Divide, Multiply, Subtract, Bring Down**. After dividing, you multiply the partial quotient by the divisor before subtracting.

**9.** $|-10| = 10$ because $-10$ is 10 units from zero. Absolute value is never negative. A common mistake is to think $|-10| = -10$, but distance from zero is always positive or zero.

Find more at
ViewMath.com/OK-Grade6

**10** Start at $(0,0)$. Move 4 left and 3 up: $(-4, 3)$. Then move 7 right: $-4 + 7 = 3$, and 5 down: $3 - 5 = -2$. The final point is $(3, -2)$.

**11** $7^2 = 49$. Then $49 - 10 = 39$.

**12** "The quotient of $m$ and 5" means $m$ divided by 5: $m \div 5$.

**13** $4^2 + 2(4) + 1 = 16 + 8 + 1 = 25$.

**14** Distribute: $4 \times 3y - 4 \times 2 = 12y - 8$.

**15** Notebooks cost $3n$ and pens cost $1 \times p = p$. Total: $3n + p$.

**16** Multiply by 8: $w = 9 \times 8 = 72$.

**17** $t \leq 30$ means 30 or less, which is "no more than 30 degrees."

**18** Closed circle means 3 is included ($\geq$ or $\leq$). Shading right means greater values: $x \geq 3$.

**19** You choose $x$ (independent), and $y$ is computed from it (dependent). $y$ increases as $x$ increases.

**20** When both bases are equal ($b_1 = b_2 = 4$), the trapezoid is actually a parallelogram. Area $= \frac{1}{2}(4 + 4)(6) = 24 = 4 \times 6$.

**21** Bottom rectangle: $10 \times 4 = 40$. Top-left rectangle: $6 \times (8 - 4) = 6 \times 4 = 24$. Total: $40 + 24 = 64$ square units.

**22** $SA = 2(10)(3) + 2(10)(7) + 2(3)(7) = 60 + 140 + 42 = 242 \ m^2$.

Find more at
ViewMath.com/OK-Grade6

**23** Data sets are described using center, spread, and shape. "Color" is not a statistical feature of data.

**24** Mean $= (4 + 6 + 8 + 10 + 12) \div 5 = 40 \div 5 = 8$.

**25** The score 7 has 5 dots — the highest count. The mode is 7.

**26** 9 values. Median is the 5th value (40). Upper half: $45, 50, 55, 60$. $Q3 = (50 + 55) \div 2 = 52.5$.

**27** A single outlier can make the range very large while the rest of the data is tightly clustered. The IQR gives a better picture of overall spread. For example, $\{1, 50, 51, 52, 53\}$ has range 52 but most values are close together.

**28** $P(5) = \dfrac{1}{6} \approx 0.1\overline{6}$, which rounds to 0.17.

**29** The values are $12, 18, 20, 24, 26, 30, 32, 38$. $Sum = 12 + 18 + 20 + 24 + 26 + 30 + 32 + 38 = 200$. Mean $= 200 \div 8 = 25$.

**30** The axes show sports (categories) and number of students (counts). The best title describes a survey of students' favorite sports.

## ✅ Practice Test 6 — Answer Key

**1** 42 miles per 3 hours   **2** B   **3** A   **4** B   **5** C   **6** B   **7** C   **8** B

**9** D   **10** C   **11** B   **12** A   **13** 16   **14** B   **15** C   **16** A   **17** $w < 50$

**18** Solutions: 0 and 5 (or any values greater than $-1$). Non-solution: $-1$ (or any value $\leq -1$)   **19** B

**20** 48 in$^2$   **21** B   **22** 348 in$^2$   **23** B   **24** C   **25** C   **26** C   **27** A   **28** D

 Find more at
ViewMath.com/OK-Grade6

 **29** C     **30** C

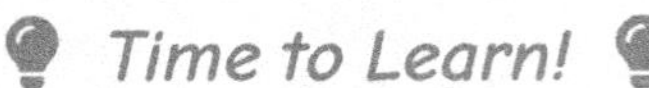

## 💡 Time to Learn! 💡

Review the explanations below, **especially for the questions you missed**.

Understanding why each answer is correct builds stronger problem-solving skills.

**Tip:** Circle any questions you got wrong, then read their explanation carefully.

## 📖 Practice Test 6 — Detailed Explanations

**1.** State both quantities with their units: 42 miles per 3 hours.

**2.** $3 \times 2 = 6$, so $5 \times 2 = 10$.

**3.** The ratio is $2 : 3$. Choice A continues with $6 : 9 = 2 : 3$. Choice B has $6 : 8$ which is not $2 : 3$.

**4.** 10% means one-tenth. $350 \div 10 = 35$.

**5.** $1\ km = 1{,}000\ m = 100{,}000\ cm$. $10 \times 100{,}000 = 1{,}000{,}000\ cm$.

**6.** A scale of 1:24 means 1 unit on the model equals 24 of the same units on the real car. The model is smaller than the real car.

**7.** $4 \div \dfrac{2}{3} = \dfrac{4}{1} \times \dfrac{3}{2} = \dfrac{12}{2} = 6$ batches.

**8.** After $36 \div 12 = 3$, bringing down 1 gives $1 \div 12 = 0$. The student skipped that zero and jumped to $12 \div 12 = 1$, writing 31 instead of 301.

Find more at
ViewMath.com/OK-Grade6

**9** The opposite of $-3$ is $3$. Both $-3$ and $3$ are 3 units from zero, but on opposite sides of the number line.

**10** The $x$-coordinate $-3$ tells you to move 3 units to the left, and the $y$-coordinate $4$ tells you to move 4 units up. Choice D swaps the coordinates.

**11** $2^3 = 8$ and $3^2 = 9$. Sum: $8 + 9 = 17$.

**12** There are 3 boxes labeled $x$ and 2 boxes labeled 1. The expression is $3x + 2$.

**13** $8^2 = 64$. Then $64 \div 4 = 16$.

**14** Distribute: $3x - 12 + 2x$. Combine: $3x + 2x = 5x$. Result: $5x + 12$.

**15** Hourly earnings: $8h$. Plus tip: $8h + 15$.

**16** Subtract 3.5: $x = 10 - 3.5 = 6.5$.

**17** "Less than 50" means strictly under 50: $w < 50$.

**18** Any number greater than $-1$ is a solution. $-1$ itself is not, since $>$ does not include the boundary.

**19** The independent variable (number of cars) goes on the $x$-axis.

**20** Each triangle: $\frac{1}{2} \times 6 \times 8 = 24$ $in^2$. Two triangles: $24 \times 2 = 48$ $in^2$.

**21** Width $= 56 \div 8 = 7$ units.

**22** $SA = 2(15)(6) + 2(15)(4) + 2(6)(4) = 180 + 120 + 48 = 348$ $in^2$.

Find more at
ViewMath.com/OK-Grade6

**23** All values are the same $(50)$. The data is perfectly symmetric, and the range is $50 - 50 = 0$, so there is no spread.

**24** Most values are between 2 and 5, but the outlier $100$ increases the sum significantly, pulling the mean up to $22.8$.

**25** Since two values share the highest frequency of 4, the data set is bimodal — it has 2 modes.

**26** A larger IQR means the middle 50% of the data covers a wider range. Class B's middle scores are more spread out.

**27** Same medians, but Athlete A's range $(1.2)$ is smaller — the times vary less, indicating more consistency.

**28** A probability of $\frac{1}{10} = 0.1$ is close to 0, so the event is unlikely to happen.

**29** Count all the leaves: $2 + 3 + 4 + 1 = 10$ data values.

**30** Add all frequencies: $5 + 8 + 3 + 4 = 20$ students.

## ✅ Practice Test 7 — Answer Key

**1** No, it is not a rate because both quantities have the same unit (pencils).

 **2** B

**3** $(2, 5)$ and $(8, 20)$ (or any equivalent pair)

**4** B

**5** 1.5 quarts; No, one 1-quart container is not enough.

 **6** B

 **7** 4

**8** A

 **9** B

 **10** B

 **11** 41

 **12** D

 **13** C

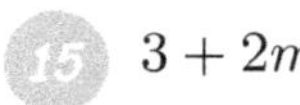 **14** A

 **15** $3 + 2m$

 **16** A

 **17** B

 **18** B

**19** C

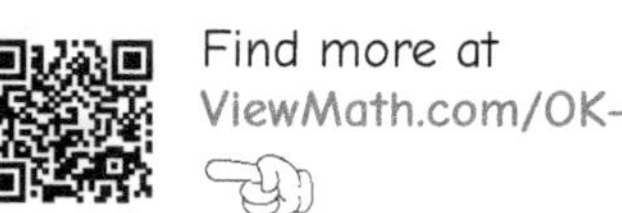

Find more at
ViewMath.com/OK-Grade6

 7 cm    B    B    Range = 8; the data is not very spread out.    B   25 No.

26 B

27 Fertilizer A: min = 8, $Q_1 = 15$, med = 24, $Q_3 = 32$, max = 42. Range = 34, IQR = 17. Fertilizer B: min = 18, $Q_1 = 25$, m...

28 B   29 B   30 B

💡 Time to Learn! 💡

Review the explanations below, **especially for the questions you missed**.

Understanding why each answer is correct builds stronger problem-solving skills.

**Tip.** Circle any questions you got wrong, then read their explanation carefully.

## 📖 Practice Test 7 — Detailed Explanations

1  A rate requires two different units. Since both amounts are measured in pencils, this is a ratio, not a rate.

2  $2 \times 4 = 8$ and $3 \times 4 = 12$, so $8 : 12$ is equivalent to $2 : 3$.

3  Ratio $= 4 : 10 = 2 : 5$. Half gives $(2, 5)$; double gives $(8, 20)$.

4  $0.15 \times 4{,}000 = 600$ children.

5  $3 \div 2 = 1.5$ quarts. Since $1.5 > 1$, one 1-quart container is not enough.

6  A scale drawing keeps the same shape as the real object but changes the size proportionally using a fixed ratio called the scale.

Find more at
ViewMath.com/OK-Grade6

(7) $2\frac{1}{2} = \frac{5}{2}$. Then $\frac{5}{2} \div \frac{5}{8} = \frac{5}{2} \times \frac{8}{5} = \frac{40}{10} = 4$ shelves.

(8) $47 \div 12 = 3$ R11; bring down $5 \to 115 \div 12 = 9$ R7; bring down $2 \to 72 \div 12 = 6$. Answer: 396. Check: $396 \times 12 = 4{,}752$.

(9) The opposite of a number is the same distance from zero on the other side of the number line. The opposite of 5 is $-5$ because both are 5 units from zero.

(10) When the $y$-coordinate is 0, the point lies on the $x$-axis. The point $(-6, 0)$ is 6 units to the left of the origin on the $x$-axis. A common mistake is thinking a negative $x$-value means the point is on the $y$-axis.

(11) $5^2 = 25$ and $4^2 = 16$. Sum: $25 + 16 = 41$.

(12) Giving away 6 means subtracting: $x - 6$.

(13) $2(9) + 2(4) = 18 + 8 = 26$.

(14) Like terms: $5p + 2p + 4p = 11p$. Constants: $3 - 1 = 2$. Result: $11p + 2$.

(15) Flat fee: \$3. Per-mile cost: $2m$. Total: $3 + 2m$.

(16) Divide both sides by 5: $n = 45 \div 5 = 9$.

(17) "Greater than 7" uses the $>$ symbol: $x > 7$.

(18) $x > 2$ does NOT include 2, so the circle should be open, not closed.

(19) Time (hours) is independent — it passes regardless. Battery level depends on time.

Find more at
ViewMath.com/OK-Grade6

20. $84 = \frac{1}{2}(10 + 14)h = \frac{1}{2}(24)h = 12h$. So $h = 84 \div 12 = 7$ cm.

21. Bottom rectangle: $8 \times 3 = 24$. Left rectangle above: $4 \times (7 - 3) = 4 \times 4 = 16$. Total: 40. This matches option B.

22. $SA = 2(4)(4) + 2(4)(9) + 2(4)(9) = 32 + 72 + 72 = 176$ cm$^2$.

23. $48 - 40 = 8$. The values are fairly close together, so the data has a small spread.

24. Mean $= (70 + 80 + 90 + 85 + 75) \div 5 = 400 \div 5 = 80$.

25. A histogram groups data into intervals. You can find the modal interval (the tallest bar), but you cannot determine the exact most-frequent value because individual values are hidden within each bin.

26. 10 values. Median $= (9 + 11) \div 2 = 10$. Lower half: $1, 3, 5, 7, 9$. $Q1 = 5$. Upper half: $11, 13, 15, 17, 19$. $Q3 = 15$. $IQR = 15 - 5 = 10$.

27. Fertilizer B's higher median shows its typical plant is taller. Its smaller IQR and range show the plant heights are less spread out. Fertilizer B is the better choice if you want tall, consistent growth.

28. Probability always ranges from 0 (impossible) to 1 (certain). It can never be negative or greater than 1.

29. The leaves for stem 5 should be written 1 3 7 9, not 3 1 7 9. Leaves must always go from least to greatest.

30. Total: $24 + 30 = 54$. Fraction of boys: $\dfrac{24}{54} = \dfrac{4}{9}$.

## ✅ Practice Test 8 — Answer Key

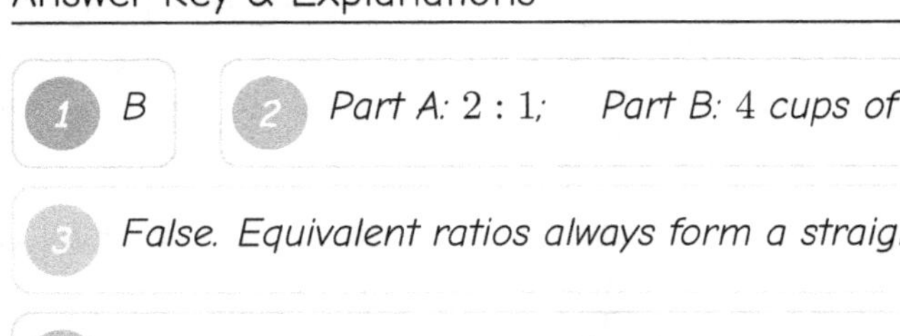

**1** B    **2** Part A: 2 : 1;    Part B: 4 cups of sugar

**3** False. Equivalent ratios always form a straight line through the origin.    **4** C

**5** Yes. 3 tons = 6,000 pounds, which is more than 5,500.    **6** D    **7** B    **8** C    **9** C

**10** Part A: Quadrant II;    Part B: $(8, 1)$, Quadrant I;    Part C: $(-8, -1)$, Quadrant III    **11** C    **12** B

**13** \$23    **14** $6x + 12$    **15** $n + 12 - 5$ or $n + 7$    **16** $a = 7$

**17** $n = 4$ is NOT a solution to $n > 4$ (since 4 is not greater than 4). $n = 4$ IS a solution to $n \geq 4$ (since 4 equals 4).

**18** C    **19** $w = 45 - 3t$; empty when $t = 15$ minutes    **20** B    **21** 24 square units    **22** C

**23** D    **24** 32    **25** Mode = 5000, Range = 4000    **26** B

**27** Mean $\approx 23.4$, Median = 12. The median is better.    **28** B    **29** 32    **30** D

---

## 💡 Time to Learn! 💡

Review the explanations below, **especially for the questions you missed.**

Understanding why each answer is correct builds stronger problem-solving skills.

**Tip:** Circle any questions you got wrong, then read their explanation carefully.

---

## 📖 Practice Test 8 — Detailed Explanations

**1** The rate compares drops to minutes: 24 drops per 4 minutes. Choice C reverses the numbers.

**2** Part A: From the graph, $(2, 1)$ shows 2 cups of flour for every 1 cup of sugar. Part B: Following the pattern, $(8, 4)$, so 4 cups of sugar.

Find more at
ViewMath.com/OK-Grade6

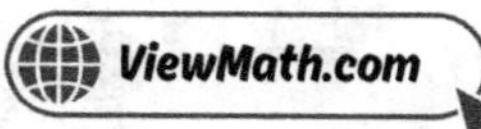

**3**   When you multiply both parts of a ratio by the same number, the points increase at a constant rate, which produces a straight line through $(0, 0)$.

**4**   $\dfrac{12}{30} = \dfrac{40}{100} = 40\%$.

**5**   $3 \times 2{,}000 = 6{,}000$ *lb.* $6{,}000 > 5{,}500$, *so yes.*

**6**   Room A actual: $(4 \times 6) \times (3 \times 6) = 24 \times 18 = 432$ *sq m.* Room B actual: $(2 \times 6) \times (3 \times 6) = 12 \times 18 = 216$ *sq m.* Total $= 432 + 216 = 648$ *sq m.*

**7**   $\dfrac{2}{7} \times \dfrac{6}{5} = \dfrac{12}{35}$. *Kai is correct.*

**8**   $73 \div 24 = 3$ R1; *bring down* $4 \to 14 \div 24 = 0$ R14; *bring down* $4 \to 144 \div 24 = 6$. *Answer:* $306$. *The zero in the tens place must not be skipped. Check:* $306 \times 24 = 7{,}344$.

**9**   The opposite of $0$ is $0$ itself. Zero is neither positive nor negative and sits right in the middle of the number line. It is the only number that is its own opposite.

**10**   Part A: $(-8, 1)$ has signs $(-, +)$, which is Quadrant II. Part B: Reflecting across the $y$-axis flips the $x$-sign: $(-8, 1) \to (8, 1)$. Signs $(+, +)$ is Quadrant I. Part C: Reflecting across the $x$-axis flips the $y$-sign: $(-8, 1) \to (-8, -1)$. Signs $(-, -)$ is Quadrant III.

**11**   $3^4 = 3 \times 3 \times 3 \times 3 = 81$. Each value is 3 times the previous one: $27 \times 3 = 81$.

**12**   The product of $4$ and $t$ is $4t$. The sum of $9$ and that product is $9 + 4t$.

**13**   $5 + 3(6) = 5 + 18 = 23$ *dollars.*

**14**   Perimeter $= 2(2x + 5) + 2(x + 1) = 4x + 10 + 2x + 2 = 6x + 12$.

Find more at
ViewMath.com/OK-Grade6

**15** *Start with $n$, add 12, subtract 5: $n + 12 - 5 = n + 7$.*

**16** $a = 105 \div 15 = 7$.

**17** *$>$ does not include equality. $\geq$ includes equality.*

**18** *$-3 \geq -3$ is true (they are equal). The other values are less than $-3$.*

**19** *Set $w = 0$: $0 = 45 - 3t$, so $3t = 45$, $t = 15$ minutes.*

**20** *The student used the slant side instead of the height. Area $= 8 \times 5 = 40$ square units.*

**21** *Base $= |5 - (-3)| = 8$. Height $= |4 - (-2)| = 6$. Area $= \frac{1}{2} \times 8 \times 6 = 24$ square units.*

**22** *A cube has 6 equal faces. SA $= 6 \times 5^2 = 6 \times 25 = 150$ in$^2$.*

**23** *Most values are between 22 and 27. The value 45 is far from the rest, making it an outlier.*

**24** *Sum $= 30 \times 4 = 120$. Fourth number $= 120 - 25 - 28 - 35 = 32$.*

**25** *5000 has frequency 6 (highest). Range $= 7000 - 3000 = 4000$.*

**26** *There are 7 values. The median is the 4th value: 9.*

**27** *Mean $= (4 + 8 + 10 + 12 + 14 + 16 + 100)/7 = 164/7 \approx 23.4$. Median $= 12$. The outlier 100 pulls the mean up. Six of seven values are $\leq 16$, so 12 is more representative.*

**28** *Numbers greater than 3: 4 and 5. That is 2 favorable outcomes. $P(\text{greater than } 3) = \dfrac{2}{5}$.*

Find more at
ViewMath.com/OK-Grade6

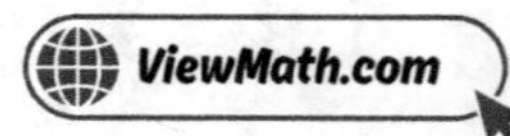

29. *The 9 values in order are* $23, 25, 28, 30, 32, 36, 37, 41, 44$. *The median (the 5th value) is* $32$.

30. *Red:* $12$, *Yellow:* $4$. *Difference:* $12 - 4 = 8$.

## 📋 Practice Test 9 — Answer Key

1. B    2. C    3. A    4. B    5. B    6. C    7. D    8. 1    9. D

10. Part A: $A = (2, 5)$, $B = (5, 1)$, $C = (2, 1)$;    Part B: $A' = (-2, 5)$, $B' = (-5, 1)$, $C' = (-2, 1)$;    Part C: Quadrant II

11. B    12. C    13. A    14. A    15. C    16. C    17. C    18. B    19. C    20. C

21. B    22. B    23. C    24. B    25. B    26. B    27. A    28. C    29. B

30. Monday, 0 *more (they are equal at 35)*

## 💡 Time to Learn! 💡

*Review the explanations below, **especially for the questions you missed.***

*Understanding why each answer is correct builds stronger problem-solving skills.*

***Tip:*** *Circle any questions you got wrong, then read their explanation carefully.*

## 📖 Practice Test 9 — Detailed Explanations

1. *Mia:* $3 \div 45 = \frac{1}{15}$ *km per min. Leo:* $5 \div 60 = \frac{1}{12}$ *km per min.* $\frac{1}{12} > \frac{1}{15}$, *so Leo is faster.*

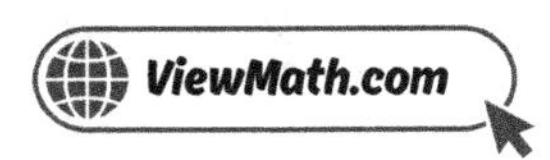

**2**  $1 \times 3 = 3$ *red, so* $4 \times 3 = 12$ *blue. Multiply both parts by 3.*

**3**  *Line A rises* 4 *for every* 1 *it moves right; Line B rises* 3. *A higher rise means steeper.*

**4**  *25% of* $\$80 = 0.25 \times 80 = \$20.$

**5**  $3 \times 8 = 24$ *pints.*

**6**  *Actual width* $= 2.5 \times 6 = 15$ *feet.*

**7**  $\dfrac{4}{5} \div \dfrac{2}{5} = \dfrac{4}{5} \times \dfrac{5}{2} = \dfrac{20}{10} = 2$ *servings.*

**8**  *After subtracting* 15, *the remainder is* 0. *Bring down* 7 *to get* 7. *Since* $7 \div 5 = 1$ *R2, the missing digit in the quotient is* 1. *The full answer is* 315.

**9**  *A number and its opposite are the same distance from zero on different sides, so they cancel each other out. For example,* $6 + (-6) = 0$ *and* $-9 + 9 = 0$.

**10**  *Part A: Reading from the grid,* $A = (2, 5)$, $B = (5, 1)$, $C = (2, 1)$. *Part B: Reflecting across the* $y$-*axis changes the sign of each* $x$-*coordinate:* $A' = (-2, 5)$, $B' = (-5, 1)$, $C' = (-2, 1)$. *Part C: All reflected vertices have negative* $x$-*coordinates and positive* $y$-*coordinates, which is the sign convention* $(-, +)$ *for Quadrant II.*

**11**  *Parentheses:* $5 - 1 = 4$. *Exponent:* $4^2 = 16$. *Add:* $16 + 7 = 23$.

**12**  *"Plus" means addition:* $n + 8$.

**13**  $6(4) - 4^2 = 24 - 16 = 8$.

**14**  $7n$ *and* $4n$ *are like terms. Add coefficients:* $7 + 4 = 11$, *so* $7n + 4n = 11n$.

Find more at
ViewMath.com/OK-Grade6

**15** $s$ represents the side length of any square. It can be any positive number.

**16** Multiply by 6: $t = 5 \times 6 = 30$.

**17** $5 > 5$ is false. 5 equals 5, but is not greater than 5. It would be a solution to $x \geq 5$.

**18** Open circle means $-2$ is NOT included ($<$ or $>$). Shading left means less than: $x < -2$.

**19** $c = 7(6) = 42$ dollars.

**20** Area $= 25 \times 14 = 350\ m^2$. Bags needed: $350 \div 50 = 7$ bags.

**21** Base $= |5 - 1| = 4$. Height $= |9 - 1| = 8$. Area $= \frac{1}{2} \times 4 \times 8 = 16$ square units.

**22** $SA = 2(6)(4) + 2(6)(3) + 2(4)(3) = 48 + 36 + 24 = 108\ cm^2$.

**23** The value 6 appears three times (the most), and the data clusters around 5–7. The peak is at 6.

**24** The data is already in order. The middle (3rd) value out of 5 is 9.

**25** Total $= 9 + 7 + 5 + 4 = 25$. Fraction choosing Basketball $= \frac{7}{25}$.

**26** Short whiskers and a narrow box indicate small range and small IQR, meaning data values are close together.

**27** Class 1's scores cluster tightly (75–85), giving a small IQR despite the outlier at 40. Class 2's even spread gives a large IQR. Class 1 is more consistent.

**28** Even numbers on a die: $2, 4, 6$. That is 3 favorable outcomes out of 6 total. $P(even) = \frac{3}{6} = \frac{1}{2}$.

Find more at
ViewMath.com/OK-Grade6

ViewMath.com

29. The stem is 5 (tens digit) and the leaf is 0 (ones digit), giving the value 50.

30. Monday: $15 + 20 = 35$. Tuesday: $25 + 10 = 35$. Both days had the same total of 35 fruits.

## ☑ Practice Test 10 — Answer Key

1. B

2. B

3. C

4. C

5. B

6. Part A: 45 km. Part B: 75 km. Part C: $45 + 75 + 60 = 180$ km.

7. $\frac{7}{4}$ or $1\frac{3}{4}$

8. 288

9. D

10. D

11. $5^2 = 25$ square units

12. B

13. $A = \$14$, $B = \$23$

14. A

15. B

16. $m = 13$

17. C

18. $x \leq 4$

19. C

20. B

21. C

22. A

23. Yes, the data is roughly symmetric.

24. A

25. B

26. $Q1 = 6.5$, $Q3 = 13.5$

27. B

28. C

29. B

30. C

## 💡 Time to Learn! 💡

Review the explanations below, **especially for the questions you missed**.

Understanding why each answer is correct builds stronger problem-solving skills.

**Tip:** Circle any questions you got wrong, then read their explanation carefully.

## 📖 Practice Test 10 — Detailed Explanations

1. From the table, the car travels 55 miles every hour. The rate is 55 miles per hour.

Find more at
ViewMath.com/OK-Grade6

2. The ratio is $2 : 50$. For 6 bags: $50 \times 3 = 150$ sq ft.

3. At \$2 per bottle, 4 bottles cost \$8, not \$10. The point $(4, 10)$ does not fit the ratio.

4. $\dfrac{45}{90} = \dfrac{1}{2} = 50\%$.

5. Converting 5 feet to inches uses the conversion ratio $\dfrac{12\ in}{1\ ft}$.

6. A to B: $3 \times 15 = 45$ km. B to C: $5 \times 15 = 75$ km. A to C: $4 \times 15 = 60$ km. Round trip $= 45 + 75 + 60 = 180$ km.

7. $\dfrac{7}{10} \times \dfrac{5}{2} = \dfrac{35}{20} = \dfrac{7}{4} = 1\dfrac{3}{4}$.

8. $60 \div 21 = 2$ R18; bring down 4 $\to$ $184 \div 21 = 8$ R16; bring down 8 $\to$ $168 \div 21 = 8$. Answer: 288. Check: $288 \times 21 = 6,048$.

9. The absolute value of 0 is 0 because zero is 0 units from itself. Zero is the only number whose absolute value is 0.

10. Quadrant IV has sign convention $(+, -)$. Since $2 > 0$ and $-5 < 0$, the point $(2, -5)$ is in Quadrant IV. A common error is confusing Quadrant IV with Quadrant II which has signs $(-, +)$.

11. The square has side length 5, so area $= 5^2 = 25$ square units.

12. Adding 12 to $h$ gives a height that is 12 inches more than Hannah's.

13. $h = 2$: $8 + 3(2) = 14$. $h = 5$: $8 + 3(5) = 23$.

14. Distribute: $7k - 21 + 5$. Combine: $-21 + 5 = -16$. Result: $7k - 16$.

Find more at
ViewMath.com/OK-Grade6

**15**  In 4 days you read 4r pages. Remaining: $100 - 4r$.

**16**  Subtract 17: $m = 30 - 17 = 13$.

**17**  "More than \$25" means greater than 25: $d > 25$.

**18**  Closed circle at 4 (included) and shading to the left (less than): $x \leq 4$.

**19**  Each dog has 4 legs: $L = 4d$.

**20**  $A = \frac{1}{2}(5 + 13)(8) = \frac{1}{2}(18)(8) = 72$ ft$^2$.

**21**  Length $= 6$, width $= 4$. Area $= 6 \times 4 = 24$ square units.

**22**  A: $2(6) + 2(20) + 2(30) = 112$. B: $6(16) = 96$. C: $2(5) + 2(8) + 2(40) = 106$. D: $2(9) + 2(18) + 2(18) = 90$. Option A is greatest.

**23**  The values are evenly spaced and balanced around the center (65). The left and right sides mirror each other.

**24**  Total students $= 15$. Median is the 8th value. In order: three 1's (3 so far), five 2's (8 so far). The 8th value is 2.

**25**  The size of the data set and whether it is numerical or categorical determine the best display. Categorical ⊠ frequency table/bar graph. Small numerical ⊠ dot plot. Large numerical ⊠ histogram.

**26**  12 values. Lower half: $5, 5, 6, 7, 8, 9$. $Q1 = (6 + 7) \div 2 = 6.5$. Upper half: $10, 10, 12, 15, 18, 20$. $Q3 = (12 + 15) \div 2 = 13.5$.

Find more at
ViewMath.com/OK-Grade6

27  Class A's higher median (80 > 75) means higher typical scores. Class A's smaller range (30 < 45) and smaller IQR (10 < 22) mean more consistent scores.

28  There are 6 equal sections and 3 of them are blue (B). $P(blue) = \dfrac{3}{6} = \dfrac{1}{2}$.

29  The stem 11 combined with leaf 5 gives the three-digit number 115.

30  The title tells us the data is about math quiz scores and specifically from Period 3, not the whole school or year.

## Well done checking your answers!

Keep practicing to strengthen your skills.

## Author's Final Note

I hope you enjoyed this book as much as I enjoyed writing it. Whether you are a student working through the material, a parent supporting your child's learning, or a teacher guiding your class, I have tried to make this book as clear and engaging as possible. I hope I have succeeded. If you have any suggestions for improvement, please let me know. I would love to hear from you.

The accuracy of calculations is very important to me. We have done our best, but I also expect that I have made some minor errors. Constant improvement is the name of the game. If you find any errors, please let me know. I will fix them in the next edition.

**For students:** Your learning journey does not end here. I have written a series of books to help you learn math. Make sure you browse through them. I especially recommend workbooks and practice tests to help you prepare for your exams.

**For parents:** Thank you for investing in your child's education. I encourage you to explore the companion resources available online to help support your child outside the classroom.

**For teachers:** Thank you for the invaluable work you do every day. I hope this book serves as a useful resource in your classroom. Feel free to reach out if you have suggestions or would like to discuss how best to use this book with your students.

I also enjoy reading your reviews. If you have a moment, please leave a review on where you found this book. It will help others find this book. If you have any questions or comments, please feel free to contact me at drNazari@ViewMath.com.

And one last thing: Remember to use online resources for additional help. I recommend using the resources on `https://ViewMath.com` You can find video lessons, practice problems, and more. You can also use the online companion for this book to track your progress and access additional resources.

Wishing all students the best in their studies, parents every success in supporting their children, and teachers continued inspiration in their classrooms!

Dr. A. Nazari

## 📖 Great Job! Keep Learning with ViewMath!

*Keep up the great work! Visit **viewmath.com/OK-Grade6** for free lessons, quizzes, and more.*

*Study Guide*

*Workbook*

*Step-by-Step*

*3 Practice Tests*

*5 Practice Tests*

*7 Practice Tests*

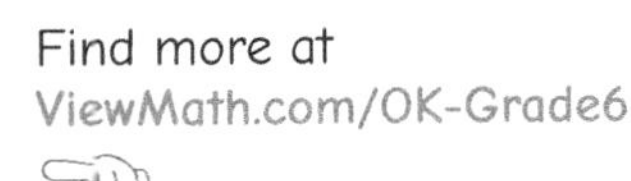
Find more at
ViewMath.com/OK-Grade6

www.ingramcontent.com/pod-product-compliance
Lightning Source LLC
Chambersburg PA
CBHW081148130726
47996CB00009B/3033